イラスト挿画／黒飛　淳
Illustration by Atsushi Kurotobi

あ.

Wilson’ s Kagoshima
Tracing the Footsteps of a Plant Hunter

ウィルソンが見た鹿児島
プラント・ハンターの足跡を追って

Tomoko Furui
古居 智子

知らない過去への旅路

2013年11月に『ウィルソンの屋久島 -100年の記憶の旅路』を上梓してから、2年余り。その間、私はウィルソンのモノクロ写真から目が離せずにいた。気がつけば、いつも画像を拡大しては飽きもせず繰り返し眺めている自分がいた。

沖縄から北海道、樺太まで合計773枚に写し取られていた自然の姿の一瞬、垣間見える当時の人々の暮らしの断片。膨大な量の写真の1枚1枚に、それぞれの地域が大正初年という時代に持っていた雰囲気が間違いなく残されていたからだ。長く沈黙の彼方にあった“物語”の気配が樹木の周りに静かに漂っている、と言ってもいいかもしれない。

いつしか私の身体はウィルソンの写真の世界の一部となり、葉の緑や幹の鳶色、花々の白や黄や紅色、樹木の横に立つ人物が身にまとっている藍染の青、神社の鳥居の目の覚めるような朱色などの色合いもほのかに感じられるようになった。ふと目を窓の外に転じれば、いつもと変わらない屋久島の山と海と空が広がっている。そのあまりにも生々しい現実の色彩の乱舞に、思わずたじろぐことさえあった。

ウィルソンが100年前に立った場所に身を置き今の風景を捉えて、そこに何が残り、何が変わってしまったのか、背景を探ってみたい。そして何よりも、その時ウィルソンが何を感じ、何を予見したかを知りたい。そんなことを夢想し始めた頃、南日本新聞社から鹿児島本土のウィルソンの写真を紹介する連載執筆の仕事をいただいた。

屋久島から始まり鹿児島市内、桜島、蒲生、霧島、川内と県内をぐるりと回り、ウィルソンが撮った写真は118枚あった。まずは、撮影場所の特定と同じアングルから現在の写真を撮ることから動き出して約1年。それは、まるでミステリー小説を読み進めるような謎解きの面白さに満ち、知らない過去への旅路へと誘われる日々であった。山の稜線を確かめ、写真の片隅に写っている建物や石壁や地形を検証し、そしてお目当ての木にたどり着いた時の興奮は言葉に表せない。里人に守られ、あるいは偶然の運で長い月日を生き延びた木もあれば、開発、災害、戦災などで株の残骸だけが残っている場所もあり、また完全にこの世から姿を消してしまった風景もあった。ウィルソンの記録を読み、地域の歴史を調べ、近隣のお年寄りに取材しては100年の時を遡って検証する作業もまた興味が尽きず、数々の新しい発見が待っていた。

これらの成果は2015年9月から半年間にわたり、鹿児島県立博物館主催の写真展で企画展示され、多くの来場者に楽しんでいただけた。

この度、1冊の本にまとめるにあたり、『ウィルソンの屋久島』では紹介することのできなかった屋久島の現在の写真や、新たに見つかった標本の写真も合わせて掲載することにした。また、本文中の植物名は一般的な名称を使用し、最終ページに学名リストを記載した。ウィルソンの写真下のキャプションは、ウィルソン自身の記述をそのまま使用している。

Journey to an Unknown Past

Two years have passed since the publication of my book, "Wilson's Yakushima," in November of 2013. Since then I could not keep my eyes away from more of the black and white photographs of Wilson's work. Without realizing it, I would find myself enlarging the digital images and gazing at them repeatedly.

Moments of nature and glimpses of people's lives throughout Japan are portrayed in the vast amount of 773 photographs. The ambiance of early 20th century Japan from each region had without a doubt been captured in them. There were hidden stories that hovered around the trees after the long silence of a century.

Gradually, I became a part of Wilson's world and started to faintly sense the color of green leaves, hazel trunks, and various colors of flowers in white, yellow and red; as well as a person clad in dyed indigo blue standing by a tree, and the bright vermilion color of the Torii of a shrine. When I gazed out the windows, there was the usual view of Yakushima's mountains, the sky and the sea that spread out. I was even surprised to realize I was surrounded by the vivid colorful world after getting used to monochrome photos.

I had the urge of placing myself where Wilson stood 100 years ago and probe the history of what has changed by documenting the present site. I wanted to know how Wilson felt and what he foresaw. It was when I started to indulge in this reverie that the Minami–Nippon Newspaper asked me to write serial stories of Wilson's expedition to Kagoshima.

Wilson took 118 photos of Kagoshima by travelling around the prefecture, from Yakushima to Kagoshima city, Kamo, Kirishima and then Sendai. I initiated my work by identifying the location of those photos and photographing the current scene from the same angle. It took almost one year and the process was like reading a mystery novel, filled with excitement of solving riddles, and simultaneously receiving invitation to journey to a past I did not know. Determining the ridgeline of mountains, verifying buildings and stone walls and geographical features which appeared in his photos, it is hard to express the joy and excitement when I finally reached the tree I was searching for. There were trees which had been protected by villagers, trees that survived through long years out of pure luck, while some others had displayed a remnant of past wreckage from natural disasters, urban development and ravages of war. Some were simply gone. I have been reading Wilson's records, studying the history of the area and interviewing the elderly to trace back the 100 years of mystery. It was a task of unfailing interest and a number of discoveries were revealed.

This project was rewarded by a six month exhibition which started from September of 2015 at the Kagoshima Prefectural Museum, which many visitors have been able to enjoy.

In compiling this book, I have included photos I could not introduce in "Wilson's Yakushima." Those are current photos of Yakushima and of Wilson's dry specimens I discovered recently. In the text I use common names, English or Japanese, for plant names. At the end of the book is a list giving both common and scientific names. The descriptions under Wilson's photographs are in his own words.

クレジット
Credit

ウィルソンの写真提供　Wilson' s Photos Credit：

ハーバード大学アーノルド樹木園

ウィルソンの標本資料提供　Wilson' s Specimen Credit：

ハーバード大学標本館

コラム
Columns

目　次
Contents

伝説のプラント・ハンター

E.H.Wilson

屋久島の「ウィルソン株」の発見で知られるアーネスト・ヘンリー・ウィルソン（1876–1930）はイギリス中西部の小さな村、チッピング・カムデンに6人兄弟の長男として生まれた。家計を助けるために、13歳から地元の園芸店に働き始めたのを皮切りに、16歳でバーミンガム植物園の庭師見習いとなり、働きながら夜学校に通い植物学の基礎を身につけた。そして21歳の時には、首都ロンドンにあるキュー王立植物園に雇用され、プラント・ハンターとして中国奥地に派遣された。派遣元となったのは、当時イギリス最大の種苗商を営んでいたヴィーチ商会で、目的はチベットとの国境近くに咲く“幻の花”、ハンカチノキ（*Davidia involucrata*）だった。

プラント・ハンターの任務は、世界の辺境の地を巡り、新種の植物を発見して持ち帰ることにある。園芸植物だけでなく、身近なところではゴムや紅茶やキニーネなどプラント・ハンターによって発見され、他地域に移植されて世界中に広がった有用植物や薬草もある。しかし、何といってもプラント・ハンティングが職業として確立したのは、大英帝国が世界に広がる植民地を舞台に植物収集戦略を繰り広げた19世紀以降のことだった。

プラント・ハンターも十人十色で、探検隊などに同行して組織的に働く人と、植物園や種苗商の依頼で単独で冒険の旅を敢行する人と、大きく分けて2種類のタイプがある。前者で日本に関係のある人物では、長崎出島を拠点に植物収集したオランダ商館医のエンゲルベルト・ケンペル、カール・ツンベリー、フィリップ・シーボルト、そして開国を求めて幕末日本にやってきたペリー艦隊のアメリカ人植物学者などがあげられる。自らの脚で歩き、自らの眼で確認し、自らの手で採集することを信条としていたウィルソンは、間違いなく後者のタイプの人だった。

ウィルソンの旅には、植物学的な意味合いだけでなく、商業的な価値のある植物の発掘などさまざまなスポンサーの思惑が絡んでいた。たった一人で現地に赴き、そこで必要な装備を整え、通訳、ガイド、ポーターや植物採集の助手を雇い、命の危険を冒して未開の地を旅しながら困難なミッションを与えられた予算で完遂することが求められた。植物学、地質学、測量学、医学の知識はもちろんのこと、現地の文化の理解、強靭な体力と精神力、交渉術、順応性、危機管理力も兼ね備え、さらにはコスト計算の能力も必要だった。

3年間の努力を経て、見事ハンカチノキの種を持ち帰ったウィルソンは、以後3回にわたる中国への探検をことごとく成功させ、“伝説のプラント・ハンター”としてのキャリアの階段を一気に登りつめた。

チッピング・カムデン本通りに面したウィルソンの生家。ドアの右上に小さなプレートが付いている
The birthplace of Wilson on the Chipping Campden main street. There is a small sign on the upper-right corner of the door.

生家から歩いて5分ほどのところにある「メモリアル・ガーデン」入り口。ウィルソンが世界中から集めた植物が植栽されている　The entrance to 'The Memorial Garden.' Plants which Wilson collected from all over the world are cultivated here.

The Legendary Plant Hunter

Earnest Henry Wilson (1876–1930), known for the discovery of the 'Wilson Stump' in Yakushima, was born at Chipping Campden, a small village in mid–western England as the eldest of six children. On finishing elementary school, he started working at a local nursery where he became interested in plants. At the age of sixteen he was employed at the Birmingham Botanical Gardens as an apprentice gardener and in the evening after work, he studied the basics of botany at a technical school. At twenty–one, he was hired by the Royal Botanical Garden at Kew in London and chosen to be sent on an expedition to China as a plant hunter. This expedition was sponsored by James Veitch and Sons, the largest nursery firm in England, in hopes of retrieving seeds of the Handkerchief Tree (*Davidia involucrata*), the 'Legendary Flower,' which was known to bloom within the borders of Tibet.

バーミンガム植物園にて。現存するもっとも若い頃（18歳）の写真（1894年）
In the Birmingham Botanical Gardens. An existing photo of Wilson's earliest days (18 years old) in 1894.

The mission for a plant hunter is to go to remote places of the world, discover unknown plants and to bring them home. Not only new species of garden plants but also more useful ones such as gum, quinine and tea were discovered by them and introduced to various other regions throughout the world. However, it was only after the 19th century, when the British began collecting plants on a large scale in their colonies around the world, that plant hunting was recognized as a profession.

Plant hunters can be divided into two types; those who accompany large organized expeditions, and those who carry out adventurous journeys on their own at the request of botanical gardens and nurseries. Of the former hunters who came to Japan were medical officers of the Dutch trading post who collected plants based in Nagasaki in the Edo period and an American botanist who came with Mathew Perry's fleet in 1853 with the purpose of opening Japan the world. Wilson was certainly the latter type of plant hunter as he firmly lived by his creed, one that required walking on his own two feet, observing with his own eyes and collecting plants with his own hands.

In Wilson's expeditions, not only was botanical value considered but also commercial value, as the discovery of valuable plants was expected by various sponsors. He had to go to the regions alone, procure the necessary equipment, and hire translators, guides, porters and assistants for collecting plants. He was required to complete his arduous mission within the given budget while risking his own life as he traveled to primitive regions. Much was demanded of him, which included risk management, cost control, negotiation skills, physical and mental strength, the understanding of local cultures, as well as having medical knowledge, surveying skills, and knowledge of geography and botany.

At the end of the expedition to primitive regions of China that lasted for three years, Wilson successfully brought back seeds of the Handkerchief tree. He then accomplished three other successful expeditions to China and quickly climbed to top of his profession as a plant hunter.

"幻の花"といわれたハンカチノキ。6月の短い間だけ花びらのような苞葉が白く色づく（北海道大学植物園にて撮影／写真提供：小野寺浩）
Davidia involucrata which was said to be 'the Legendary Flower.' During a short time in June, the bract that looks like a petal turns white.
At Hokkaido University Botanic Garden. Photo courtesy of Hiroshi Onodera.

ウィルソンの写真の魅力

ウィルソンは1899年から1911年にかけて通算9年余、4回にわたって中国奥地を歩き、その後1914年から1919年にかけて通算3年、2回の日本探検では屋久島、沖縄、小笠原諸島といった島々と本土、そして当時日本帝国の領土であったサハリン（樺太）、台湾、朝鮮（韓国）を回った。その間に1000種以上の植物を西洋に紹介し、2488枚の写真を撮った。

中国では険しい渓谷や危険な河川を前に、命がけで新種の植物を探し求めた。しかし、最後の旅で崖崩れに遭い、生死の境をさまよった。そんな過酷な中国の旅と比べ、町から町へ動脈のように延びる鉄道路線を利用した日本の旅は「まったく性格の異なる旅で、私にとっては休暇旅行のようなものである」とウィルソンは述べている。日本では、新種植物の発見というよりも主に針葉樹とサクラ、そして日本の栽培技術や園芸文化の研究といった植物学的な調査が目的とされた。右脚を負傷したため、一時は探検を断念したウィルソンにフィールドに戻る機会を与えたのは、スポンサーとなったアメリカ、ハーバード大学アーノルド樹木園園長のサージェント教授だった。

ウィルソンは野に解き放たれた狩人のごとく、日本においてもことさら他のプラント・ハンターが手を触れていない地域を選んで足を踏み入れ、道なき道を歩き、山に登り、休むことなく列島を駆け回って実際にいくつか新しい発見もしている。そして植物採集、乾燥標本の作成とともに、カメラで植物や風景を実写するという素晴らしい業績を残した。

ウィルソンの仕事の特徴は、何といってもこの写真撮影にある。19世紀末にカメラが実用化されるまで、プラント・ハンターには必ず植物画を作成する画家が付き添っていた。しかし、彼らが描いた画は標本をモデルに彩色を施した“細密画”の域を出ていない。ウィルソンはカメラという最新の機器を使って、植物が生育している自然の状態をそのまま画像として記録した。

父親の旅に同行した娘のミリュエルは、こんな言葉を記している。

「木の周りを歩き、異なる視点から何度も研究し、幹、枝の構成、角度など満足するまで入念に確認して最適の光の時を選んだ。この儀式は毎回、毎回繰り返された」

時にはロープで体を縛って、細い枝によじ登るという危険な撮影にも臆さずに挑んだという。

こうして生まれた完成度の高い作品は、美術品としての側面も高く評価されている。また、樹木の背景に山並み、田園風景や建造物などを入れ込むだけでなく、風景そのものを捉えた写真も少なくない。ウィルソンは目の前にあるものを記録する道具としてのカメラの価値を熟知し、写真に“物語”を与え、時代そのものの息づかいをも写し取ろうとしたのだ。

図書室、資料庫があるアーノルド樹木園の中枢施設、フェンネル管理棟（現在）
Hunnewell Administration Building, the central facility of the Arnold Arboretum with library and the archive (Present).

ウィルソンが佇んでいたフェンネル管理棟玄関前の階段の風景は、今も変わらない　1922-09-25
The scenery of the stairs in front of the Hunnewell Administration Building. The entrance where Wilson stood has not changed.

The Charms of Wilson's Photographs

Wilson traveled four times deep inside China, which took more than 9 years in total, between 1899 and 1911. Later he conducted two expeditions to Japan for a total of 3 years from 1914 to 1919. He traveled the Japanese mainland and islands such as Yakushima, Okinawa and Ogasawara as well as Sakhalin, Taiwan and Korea, which were territories of the Japanese Empire during those years. He introduced more than 1000 species of plants to the Western world and took 2488 photographs.

In China, he was confronted by steep ravines and dangerous rivers where he literally risked his life in search of new plants. In comparison to such extreme trips, the journey in Japan utilized a railway system that extended from town to town like an artery. Wilson wrote to his mother that this trip would be "one long holiday for me." The purpose of his trip to Japan was mainly a botanical investigation of conifer and Cherry trees, and to gain insight into Japanese cultivation methods and gardening culture. This gentler expedition, at the request of Professor Sargent of the Arnold Arboretum, was in consideration of Wilson's physical condition. He was severely injured in a sudden landslide on his last expedition in China.

However, Wilson was like a hunter let loose in the field of Japan, as he selected locations that other plant hunters had not explored. He cut his way through untouched wilds, climbed mountains, and ran about the Japanese archipelago without resting while making new discoveries. In addition to collecting plants and producing dry specimens, he recorded his remarkable achievements by taking photographs of plants and scenery everywhere he went.

Wilson's splendid work of taking photographs was undoubtedly the most noteworthy part of his career. Until the camera was put into practical use at the end of the 19th century, plant hunters were always accompanied by a painter who produced images of plants. However, the images they drew were merely colored models of the specimen. Wilson used the latest device called the camera to capture the natural state of plants, in their settings, in highly detailed images.

His daughter, Muriel has stated the following; "He would walk around a tree and study it from different points of view over and over again and check the structure and angle of trunk and branches thoroughly until he was satisfied. And then, he chose the best moment for lighting. This ritual was repeated every time." He was known occasionally to challenge the dangerous task of tying himself up with a rope to climb thin branches for a shooting.

With a high degree of perfection, his work was greatly appreciated as fine art. He not only took photographs of trees in their natural setting but included buildings and scenes of rural fields in the background. Wilson acknowledged the value of the camera as a tool that would record things we see in front of us, but also attempted to portray the 'time' itself by telling a story.

フェンネル管理棟内、ウィルソンの書斎（現在）
Wilson's study in the Hunnewell Administration Building (Present).

左と同じ部屋　1924-11-05
The same room as the left photo.

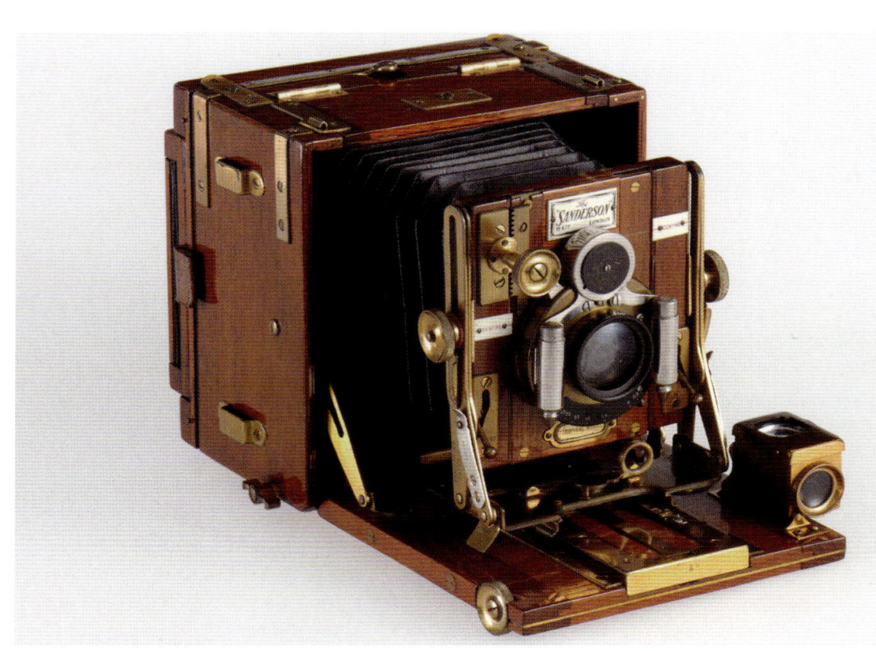

ウィルソンが日本での撮影に使用したものと同型の「サンダーソン・カメラ」(© 日本カメラ博物館 JCII CAMERA MUSEUM)
'The Sanderson Camera' of the same type that Wilson used for photography in Japan.

ザ・サンダーソン・カメラ

常にウィルソンの良き理解者であったアーノルド樹木園のサージェント教授は、価格に関係なく最高品質のカメラを購入するようアドバイスした。ウィルソンはこの言葉に従い、建造物を撮影するためにイギリスのサンダーソン社が開発した最新鋭のフィールド・カメラを手に入れ、プロのカメラマンに習って熱心に撮影の練習を重ねた。

上面のフラップが開きレンズが上がる仕組みになっているため広い遠景も歪みなく撮影でき、高品質な大量のデータをガラス乾板に記録できるこのカメラは、フィールドで背の高い樹木のイメージを捉えるには適していた。しかし露光時間は5-8分と長く、風のない晴天の日でも1日最大15枚撮影するのが限度であった。また、箱型蛇腹式の本体を収納する3つの大きな箱に頑丈な木製の三脚、それに多数の壊れやすいガラスの板とそのホルダーを入れた重いケースを何箱も山中や僻地に運び込まねばならなかった。

撮影したガラス乾板は慎重に荷造りして船便でロンドンに送り、後日すべてを自分の管理の下で現像させたという。

なぜ、それまでして苦労を重ねて写真を撮ろうとするのか、新聞記者に尋ねられてウィルソンはこう答えている。

「私たちが今、記録を残さなかったら、100年後にはその多くは完全に消えてしまうだろう」

フィールド植物学者であったウィルソンは、写真という記録媒体を通して、100年後の私たちに貴重なメッセージを伝えているのである。

ウィルソンが画像を写したガラス乾板（15.5cm × 20cm）
The glass photographic negative which Wilson recorded.

上のガラス乾板を現像したもの
The photo that was developed from the glass plate.

The Sanderson Camera

Professor Sargent of the Arnold Arboretum instructed Wilson to purchase a camera of the highest quality regardless of the cost. Wilson followed his advice by purchasing a cutting edge Field Camera from Sanderson & Company in England which was developed for the purpose of taking photographs of tall buildings. He learned how to take photographs from a professional photographer and devoted his time in building the necessary skills.

Because the front lens was designed to tilt in order to correct for perspective distortion, it was ideal for photographing towering trees in the field. The large glass plate negatives recorded a massive amount of high quality data. However, the exposure time was a long 5 to 8 minutes, so even in broad daylight without wind, photographing a maximum of 15 images a day was the limit. In addition, they had to carry three large boxes that contained the main camera, a sturdy wooden tripod, and a number of heavy cases of fragile glass photographic plates and plate holders into the mountains and remote regions.

Moreover, he had to pack each one of the recorded glass photographic plates carefully, and ship them to London to be developed under his supervision.

A newspaper reporter once asked, “Why do you continue to go through the trouble and still insist on taking photographs ? ”

Wilson replied, **“If we do not get such records of them, a hundred years hence many will have disappeared entirely. ”**

Wilson, who was a field botanist, left a valuable message for people today, after 100 years, through the medium of photography.

ウィルソンの鹿児島

38歳の誕生日を目の前にしたウィルソンが初めて横浜の港に降り立ったのは、今から102年前の1914（大正3）年2月3日のことだった。真っ先に目指したのは、ちょうど3週間前の1月12日に桜島大噴火を経験したばかりの鹿児島だった。

1859（安政6）年に長い鎖国の世が開け、横浜、長崎、箱館（函館）の港から貿易が始まると、未知の植物の宝庫、日本を目指してロバート・フォーチュン、カール・マキシモヴィッチといった著名なプラント・ハンターたちが堰を切ったように各国から押し寄せた。しかし当時、外国人の国内旅行は禁じられ、行動範囲も限定されていたため、日本人の採集人や植木屋などを通して植物を手に入れることが多かったという。加えて攘夷運動の高まりもあり、あえて危険を冒してまで薩摩の国を訪れる外国人はいなかった。その後、植物採集の興味は中国に移り、来日するプラント・ハンターの数も減った。1877（明治10）年に維新後の日本を訪れたチャールズ・マリーズなどは耐寒性のある植物を求めてか、もっぱら東北から北海道を巡っている。

花の開花に合わせて南から日本列島を北上する計画であったと思われるが、ウィルソンにとっては鹿児島のこの“空白性”こそが、訪れる価値のあるものだった。また鹿児島での日々は、その後の旅の原点となった。言い換えれば、日本各地の植物や風景を比較検証する際の基準の地として、ウィルソンが絶えず振り返ることになる重要なフィールドだった。

新橋駅から列車を乗り継ぎ、鹿児島市内で1泊すると、蒸気船に乗ってまずは、種子島経由で屋久島に上陸、2月18日から9日間にわたり山や里を探索した。途中で発見した大株が、「ウィルソン株」と名付けられ、今や屋久島を代表する観光スポットになっている。

鹿児島本土に戻ってからは、3月2日から16日までの15日間、県内を巡った。市内で旅装を解くと、まず訪れたのは城山だった。次いで重富、蒲生を経て霧島山に入り、その後は再び市内を経由して西に舵をとり川内の神社に足を運んだ。最後は桜島に上陸し、その翌日に島津公爵の庭園（仙巌園）を訪れ、合計118枚の写真と288点の標本を手に鹿児島を後にした。

こうしてウィルソンの足跡を追ってみると、短い時間で実に効率的に県内の名所、銘木を見て回っていることに驚く。この旅の功績の陰には有能な案内役の存在があったに違いない。改めて102年前の写真をじっくりと見てみると、何枚もの写真に樹木と一緒に写っているひとりの人物が浮かびあがる。ウィルソンが「最も魅力的で熱心な旅の友」とボストン宛の手紙に記したこの青年は、当時鹿児島大林区署（現九州森林管理局）で林業技師として働いていた三好哲男だった。

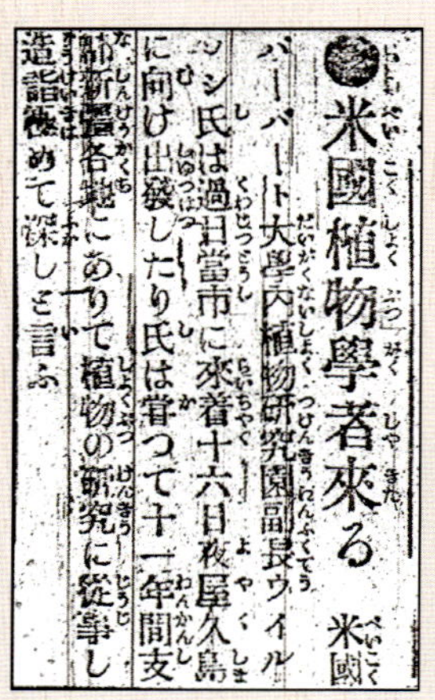

米國植物學者來る

米國ハーバート大學内植物研究園副長ウイルソン氏は過日當市に來着十六日夜屋久島に向け出發したり氏は嘗つて十一年間支那新疆各地にありて植物の研究に從事し造詣極めて深しと言ふ

ウィルソン来鹿を伝える当時の新聞（鹿児島新聞大正3年2月18日）
A newspaper article that reported Wilson's visit to Kagoshima. (Kagoshima Newspaper, February 18, 1914)

鹿児島にて撮影、三好哲男（写真右）
1917-03-17
Tetsuo Miyoshi (Right) at Kagoshima.

Wilson's Kagoshima

Before his 38th birthday on February 15, Wilson landed at Yokohama port on February 3, 1914. The first location he headed for was Kagoshima prefecture, where Sakurajima, a symbolic volcano in Kagoshima city, had just experienced an historic eruption on January 12, just 3 weeks earlier.

After the long, closed–door policy had ended in 1859, foreign trade began from the ports of Yokohama, Nagasaki and Hakodate. Many prominent plant hunters, such as Robert Fortune and Carl Johann Maximowicz flocked eagerly from various countries to Japan, a place known as a gold mine of undiscovered plants. However, at that time, domestic travel was prohibited to foreigners and the range of activities was limited. It was not uncommon for the plant hunters to procure plants through Japanese gardeners and plant collectors. In addition, in the wake of chaos that followed the Anti–foreigner Movement, few foreigners would risk their life on a mere visit to Kagoshima. After that, the zeal for collecting plants was concentrated in China and plant hunters who visited Japan had decreased in number. Charles Maries, who came to Japan after the Meiji Restoration in 1877, for example, had traveled exclusively to the north area, perhaps in search of hardy plants.

It seemed Wilson had first intended on traveling northward in Japan to follow the Cherry blossoms, but perhaps the fact that Kagoshima was still a 'blank area' which Western plant hunters had never explored was enticing to him. Thus Kagoshima became an important reference point as he compared its plants and scenery to other parts of Japan.

He took trains from Tokyo to Kagoshima, spent one night in Kagoshima city and took a steam boat to Yakushima via Tanegashima, then delved into the mountains and villages for 9 days from February 18. During his excursion he discovered a big stump, which was later called the 'Wilson Stump.' It has become one of the main tourist spots that represent Yakushima today.

Following his return to the mainland, he spent 15 days until March 16, traveling around Kagoshima prefecture, both its main city and surrounding area. He left Kagoshima after investigating Sakurajima with a total of 118 photographs and 288 dry specimens.

As I traced Wilson's footsteps, I was amazed how effectively he had visited well known places and treasured trees around Kagoshima prefecture in such a short period of time. I came to the conclusion that there must have been a very capable dragoman behind his achievements. As I carefully gazed at Wilson's photographs, I noticed a person standing by a tree in several of them. In a letter addressed to Boston, Wilson stated this young man was "a most charming and enthusiastic traveling companion." I was able to discover that it was Tetsuo Miyoshi who worked as a forestry engineer in the Kagoshima Forestry Office.

屋久島 Yakushima 1914-02-23
スギ *Cryptomeria japonica* D.Don

ウィルソンの写真のなかの三好哲男
Tetsuo Miyoshi in Wilson's photos.

霧島 Kirishima
1914-03-07
モミ
Abies firma S. & Z.

日本帝国の鉄道路線図。ウィルソンが持参していたもので、訪れた場所と移動経路に青鉛筆で印がつけられている
The map of Japanese Imperial Government Railways.
Wilson brought this map with him. The places he visited and route he traveled were marked with blue pencil.

No. 321
PASSPORT.
By His Britannic Majesty's
Vice Consul
at
Boston Massachusetts U.S.A
Ernest Henry Wilson
accompanied by his wife
DESCRIPTION OF BEARER
PHOTOGRAPH OF BEARER
RENEWALS
SIGNATURE OF BEARER

入国時のパスポート
His passport at the time of entry to Japan.

一人娘のミリュエル（当時８歳）
Muriel, Wilson's only daughter (8 years old) .

日本の旅で、初めて妻と娘を伴った。ウィルソンは探検の途中で東京や軽井沢のホテルに滞在していた２人と合流し、家族の時間をもった
For the first time, on his trip to Japan Wilson brought his wife and daughter. They stayed at hotels in Tokyo while he traveled, but he would join them when back from his exploration.

日本探検では主に鉄道を使って移動した。プラットフォームに立つウィルソン（左）（1917–1919 年）
He moved mainly by rail for the exploration of Japan. Wilson standing at the platform (left) .

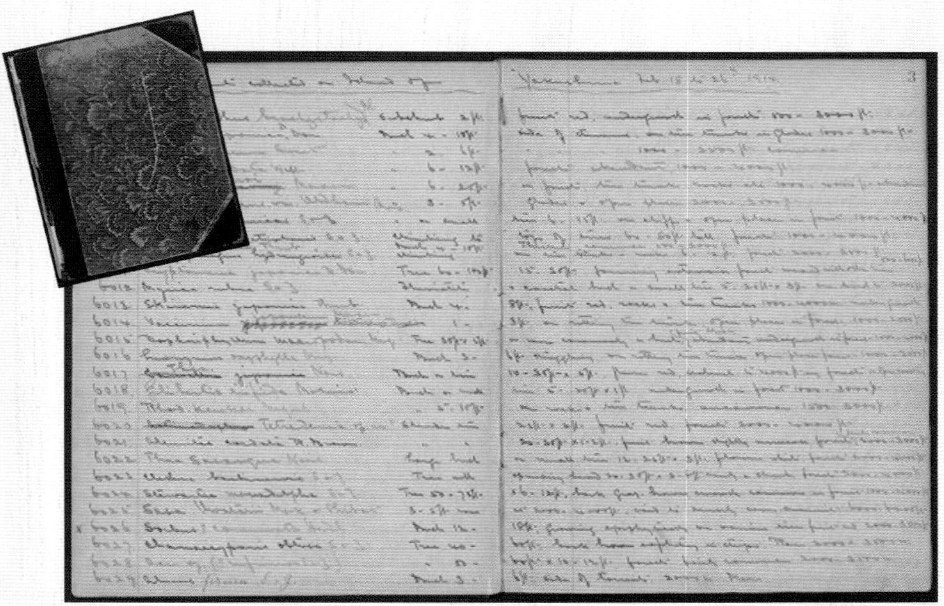

ウィルソン手書きのフィールド・ノート（1914 年２月 –1915 年１月）。最初のページ右上に「屋久島 1914 年２月 18 日」の記述がある
Field notebook of Wilson with his handwriting (February, 1914–January, 1915) . The description of 'Yakushima February 18th, 1914' is seen at the top-right corner.

column ウィルソンの遺産　Wilson's Legacy

ウィルソンは生涯5000種以上の新しい植物を紹介したとされるが、その正確な数はいまだ誰にもわからない。ヨーロッパやアメリカの個人の庭、植物園、景観のなかには少なくとも1種類は必ずウィルソンがもたらした植物が含まれるといわれている。

文才にも恵まれた人で、見聞したことを日記や手紙に書き残すだけでなく、研究者向けの論文の他に一般読者向けの本も合わせて12冊ほど出版し、園芸雑誌や植物学の定期刊行物にも数えきれないほどの寄稿をしている。スライド化した写真をもとに各地で講演を行い、ラジオ番組にも登場して人気を博した。欧米諸国の一般人を対象にガーデニングの楽しみを広げたことは、ウィルソンの大きな功績のひとつであろう。

また、盆栽やサクラの品種改良などに見られる日本の秀でた園芸文化も海外に紹介し、植物を介しての東西交流に貢献した。国際的な園芸貿易の発達に伴い、英文パンフレットを作って積極的に海外に販路を広げる日本の植木会社と親しく交わった。さらに、旅の途中で出会った日本人の植物学者や研究者とも情報交換などを通して積極的に交流を深めた。

頑迷な学究の徒でもなければ、孤高の冒険野郎でもなかった。ウィルソンはフィールドにあっては現地の人々と人間的な付き合いを求め、そして仕事の成果を惜しみなく広く一般に提供し続けた。その情熱の源には、常に自然への深い尊敬と愛情があったといえる。

1930年10月、ボストン郊外で車の車輪が枯れ葉でスリップして崖から転落するという痛ましい事故に遭遇し、伝説のプラント・ハンターは54歳でこの世に別れを告げた。後世に多くの遺産を遺した短くも鮮烈なナチュラリストの一生だった。

ウィルソンは横浜や大阪の園芸店を訪れ、親交を深めた（1917–1919年）
Wilson visited the nurseries of Yokohama and Osaka and built friendships with them.

Wilson is said to have introduced more than 5000 species of new plants in his lifetime, but no one really knows the exact number. Some say that at least one kind of plant brought by Wilson to the regions of Europe and America can be seen in any home garden, botanical garden, or general landscape.

He also had a great aptitude for literary work. Besides the experiences and observations noted in his diary and letters, he wrote as many as 12 books that were intended for both scholars and general readers. He made innumerable contributions to garden magazines and periodical journals in the field of botany. Furthermore, he gave lectures in various places with slides made from his photographs, and appeared on a radio program that won popularity. Perhaps one of his greatest achievements is the result of having spread the pleasure of gardening to ordinary people in America and Europe.

In addition he introduced Japanese gardening culture, such as the art of Bonsai and the breeding techniques of Cherries to the west, while contributing to East and West cultural exchange through plants. He formed friendships with Japanese nurseries which had expanded the markets abroad by proactively using English brochures and exchanged information with Japanese botanists and researchers.

He was neither an obstinate scholar, nor an isolated daredevil. When he was in the field, he sought human connection with local people and continued to generously provide the results of his work to the public. I think the source of his passion came from his deep respect and love for nature.

In October, 1930, Wilson died in a tragic car accident in Boston that had resulted from slipping on fallen leaves and plunging off a cliff. The legendary plant hunter left this world at the age 54. It was a short but vibrant life of a naturalist who left many legacies to future generations.

第1章
屋久島
Scene I　Yakushima

スギ　*Cryptomeria japonica* D.Don.　1914-02-23

太古の島　ヤクスギの森

プラント・ハンターの嗅覚だろうか。情報を得るために訪れた東京帝国大学（現東京大学）小石川植物園で、鹿児島南西の洋上に浮かぶ島の名を耳にした瞬間に、ウィルソンの日本での最初の一歩が定まった。

「屋久島に行こうと思う。ぜひとも、私自身の目で見てみたい」

2月17日、噴煙をあげる桜島を背に鹿児島港を出た蒸気船「大川丸」が屋久島宮之浦港に接近するにつれ、海から空に向かって聳え立つ山並みが徐々に視界を埋めていった。ウィルソンが自らの目で確かめたいと願ったのはヤクスギの天然林であった。

翌日、ガイドの手配がつくとすぐに山に分け入り、標高780m付近をベースに、積雪で行く手が拒まれる1030mあたりまで歩き回った。そこで見たのは普通の杉の約5-6倍、多いものだと20倍ともいわれる樹脂含有量をもち、そのせいで樹齢数千年という驚異的な長寿を保っている巨大スギが君臨する太古の森であった。従者のように立ち並ぶ針葉樹のツガやモミ、広葉樹のヤマグルマやヒメシャラなどの大木。そして樹木や岩に着生するさまざまなツタ、コケや地衣類。年間降水量4,000mmから8,000mmという特殊な自然環境が生み出す、湿潤にして生命力に満ちた神秘の宇宙がそこに広がっていた。

ヤクスギは古くから建築用材として珍重され、藩政時代には屋根を葺く平木にして搬出されるほか、遠くは京都や奈良の寺院などの建造物の需要に給された。森の中には3-4mの高さの切り株がたくさん点在し、その上にさまざまな世代の樹木が新しい命を育んでいる。

「外国人にしろ、日本人にしろ、過去に誰もこれらの植物を研究する機会を楽しんだ人間の記録はない」

興奮した口調でこう書き残したウィルソンだが、実はさかのぼること5年前、1909（明治42）年9月にこの誰も注意を払うことのなかった南の離島を訪れた一人の日本人植物学者がいた。後に日本の“植物学の父”と呼ばれた牧野富太郎である。牧野は屋久島の最高峰宮之浦岳（1936m）に登り、山中で4泊してカンツワブキ、ヤクシマリンドウ、ヒメヒサカキなどの固有種を採集した。小学校を2年で退学して独学で道を究めた牧野と、似通った背景を持つウィルソンとは共感することが多かったのかもしれない。双方の著述でお互いの仕事を意識し、尊敬し合っている。

ともあれ、屋久島の山を探索した最初の西洋人はウィルソンであることには間違いない。帰国後に出版した論文『日本の針葉樹』で、ウィルソンは初めて「屋久島」の名を世界に紹介した。

スギ　1914-02-20
Cryptomeria japonica D.Don.

モミ　1914-02-23
Abies firma S. & Z.

イスノキ　1914-02-19
Distylium racemosum S. & Z.

An Ancient Island; the Forest of Giant Trees

Could it be because of his intuition as a plant hunter ? Wilson's first step on his expedition to Japan was decided the moment he heard of an island of virgin forests that floated on the sea south of Kagoshima, while visiting the Imperial Botanical Garden in Tokyo.

He wrote, "I go to the island of Yakushima and I am all eager to see them for myself."

On February 17, a steamboat departed Kagoshima Port, leaving Sakurajima behind in the background as it spewed out volcanic smoke. As the boat approached Miyanoura Port on Yakushima, the landscape of mountains, which soared into the sky, gradually filled Wilson's view. What he eagerly wanted to see for himself was the primeval forest of Yaku–cedar (*Cryptomeria japonica*).

The next day, as soon as arrangements for guides were made, he pushed his way into the mountains. At approximately 780m above sea level he set up base camp. From there he explored the vicinity until they came up to the altitude of 1030m where snow covered ground hindered them from going on. What he saw there was an ancient forest dominated by gigantic Yaku–cedar with an amazing longevity of several thousand years, due to its resin content being 5 to 20 times more than a usual Cedar. There were coniferous Tsuga (Japanese Hemlock) and Momi (Japanese Fir) and large broadleaf trees such as the Wheel Tree (Trochodendron) and Japanese Stewartia who stood like valets. Various epiphytic ivy, moss and lichen, which grow on trees and rocks, could also be seen. Sprung from dampness caused by an annual precipitation of 4,000 to 8,000 mm, a mystical world filled with vitality existed before Wilson's eyes.

Yaku–cedar has been highly valued as architectural material. It was primarily cut up into roof shingles which were shipped as far as Kyoto and Nara to meet demand for building temples there during the Edo period. Many stumps of 3 to 4 m in height were scattered around the forests, with various generations of trees and plants growing from their tops.

"There is no record of anyone, foreigner or native, having previously enjoyed such opportunity to study these plants," Wilson wrote in an excited voice. However, in September of 1909, five years before Wilson's survey, a Japanese botanist had visited this island. It was Tomitaro Makino who would later be called the 'Father of Japanese botany.' Makino climbed the highest peak of Mt. Miyanoura (1,936m) and spent four nights in the mountains, collecting endemic species such as Farfugium, Yakushima Gentian and Yakushima Eurya. Makino, who left elementary school as early as second grade and mastered his way by self–education, must have had many things in common with Wilson who shared a similar background. It can be read from their writings that they were conscious of each other's work and respected them.

Regardless of Makino's visit, there is no mistake that Wilson was the first Westerner who searched the mountains of Yakushima. After his return home, Wilson introduced the name of 'Yakushima' to the world for the first time in his book, "The Conifers and Taxads of Japan."

ツガ *Tsuga Sieboldii* Carr. 1914–02–19

E.H. ウィルソン著
『日本の針葉樹』
1916-12-30

"The Conifers and Taxads of Japan" by Ernest Henry Wilson, Dec. 30, 1916.

スギ　*Cryptomeria japonica* D.Don.　1914-02-20

幹周 9.1m。ウィルソンが屋久島山中で出逢った最大の屋久杉　標高 1030m
Girth 30ft. The largest known wild specimen. alt. 1030m

採集日 Collection Date = **1914-02-19**

日本の固有種であるスギは屋久島がその南限で、樹齢1000年以上の古樹のみが、屋久島では通称ヤクスギと呼ばれている。写真は、ウィルソンが屋久島の山の中で出逢った巨大なヤクスギの幹周を計測しているところ。幹には深い割れ目が入り、樹皮は赤茶色で、長い年月風雪にさらされた部分は灰色になっている。

ヤクスギは通常、標高650m以上の暖温帯から生えるが、撮影地はアカシデ、ハリギリやユズリハなどが生育する高標高の冷温帯だった。この辺りでは、樹高40m、幹周5-10m級の巨大スギがたくさん見られたが、戦後の大量伐採でその多くは姿を消した。

Cryptomeria japonica is an endemic Cedar of Japan, and its southern limits are Yakushima. Only aged trees that are more than 1,000 years old are commonly called Yaku–cedar in Yakushima. In this photograph, Wilson is measuring the trunk circumference of an enormous Yaku–cedar that he came across in the mountains of Yakushima. A deep crack can be seen on the trunk. The bark is reddish brown and the part exposed to wind and snow for a long time has become gray.

Yaku–cedar normally grows in a warm temperate zone at an altitude of more than 650m. However, the site of Wilson's photograph was in a cool temperate zone at high altitude, where Akashide, Harigiri and Yuzuriha grow. Many large Yaku–cedars with a height of 40m and trunk circumference of 5–10m had been seen in this area, but most of them disappeared due to postwar mass felling.

ヒメシャラ　*Stewartia monadelpha* S. & Z.　1914-02-20

滑らかで薄茶色の樹皮をもつ。幹周 3.2m 標高 780m

Bark smooth, pale brown. Trunk girth 10.5ft. alt. 780m

採集日 Collection Date = 1914-02-20

ウィルソンが“森の貴婦人”と呼んだ、光沢のある明るい褐色の幹が特徴的な落葉樹のヒメシャラ。その美しい立ち姿は、緑の針葉樹のなかでひときわ目をひいた。写真の奥にはヤマグルマの姿も見える。

フィールド・ノートの記録によると、ウィルソンは標高300-1500mのところで、樹高15-25m、幹周2-4m級の個体に出逢っている。この日、ヒメシャラの剥がれた樹皮と葉のない実付きの標本を作成した。

It was the distinctive Japanese Stewartia (*Stewartia monadelpha*), a deciduous tree which possessed a bright brown trunk, that Wilson called 'the lady of the forest.' It's beautiful straight posture was conspicuous among the green conifers. The figure of Wheel Tree (*Trochodendron aralioides*) can be seen in the back of the photograph.

According to the record from his field notes, Wilson encountered several distinctive trees with a height of 15–25m and a trunk circumference of 2–4m at altitudes of 300–1300m. On that same day, he prepared a specimen of Stewartia bark and branches with bud and fruit.

ヤマグルマ　*Trochodendron aralioides* S. & Z.　1914-02-22

樹高 16.8m 幹周 7.2m 樹冠 15.2m　右にスギが見える　標高 750m
Height 55ft. Circumference 23.5ft. crown 50ft. through. Cryptomeria to right. alt. 750m

採集日 Collection Date = **1914-02-24**

ヤクスギの切り株の上で、大きく枝を広げて生育するヤマグルマ。樹皮はびっしりと苔むしていた。写真の右に別のスギが見える。「この森でスギとヤマグルマは、お互いに奇妙な愛情で結ばれているに違いない」とウィルソンはノートに記している。

ヤマグルマはスギ、モミ、ツガなどの幹上で芽生えて取り付き、根を地上に下ろす。時には取り付いた木を殺してしまうこともある絞殺木である。森の中では、生存のための闘いが終わることなく続いている。

A Wheel Tree (Trochodendron) spreads its branches and grows from the top of a moss–covered stump of a Yaku–cedar (Cryptomeria). The bark is fully covered by moss and another Cedar can be seen on the right of the photograph. "The *Trochodendron aralioides* and *Cryptomeria japonica* in this forest must be bounded by a strange affection for one another," Wilson wrote in his notes.

The Wheel Tree germinates in the branches and stems of Cedar, Momi (Japanese Fir) or Tsuga (Japanese Hemlock), and sends their roots down to the ground. At times, they are known as the 'Strangulation tree' that strangles the tree it grows on. In the forest, the fight to survival continues endlessly.

リョウブ　*Clethra barbinervis* S. & Z.　1914-02-24

樹高 10.7m 幹周 1.5m 標高 800m　左にモミ

Height 35ft. Circumference 5ft. alt. 800m. *Abies firma* S. & Z. to left.

採集日 Collection Date = 1914-02-20

大きく傾いたリョウブと、奇妙なカーブを描く幹の下で彫像のように立ち尽くす若者の姿。そのコントラストが、絵画的に見える１枚である。赤茶色の樹肌をもつリョウブは屋久島が分布の南限で、ウィルソンは樹高 6–10m、幹周 1–2m の個体をところどころに見たと記している。

現在も標高 500–1600m 付近の登山道脇によく見られる。

A young man standing still like a statue under a strangely curved reddish brown trunk of a Japanese Clethra growing diagonally. This photograph contrasts them in a picturesque way. Yakushima is the southern limit of distribution for Japanese Clethra. Wilson wrote that he saw examples of 6–10m in height with a circumference of 1–2m in several places.

Even today, they are commonly seen along mountain trails between altitudes of 500–1600m.

100年目の奇跡

「親父……」

牧市助はモノクロの「ウィルソン株」の写真を見た瞬間、そう絶句すると目を潤ませた。

ひょっとして写真のなかの人物を特定できるかもしれない。そんな淡い期待を抱いて屋久島で開いた写真展「ウィルソンの屋久島」の最終日夕方、妻に誘われて会場を訪れた市助は、ウィルソンと父親の関わりを初めて公にした。

ウィルソンは樹木の写真を撮る時に大きさを示す尺度として、必ずといっていいほど木の近くに人を配している。屋久島の奥岳、標高1030mのところで見つけたヤクスギの切り株は幹周15.2mと稀に見る巨大さであったためか、上下左右に４人の人物を置いた。森林局の職員と思われる左下の制服姿の男性を除く３人は、綿入れの着物に脚絆、地下足袋という装いであることから、地元のガイドであろうと想像できた。市助が指さしていたのはそのなかの左上、童顔の男性である。名前は牧次郎助、市助の父親だった。屋久島での写真ではいちばん多く登場し、明らかにウィルソンが信頼を寄せ、心を許した若者であったと思われる。

「中国でさえ、このような豊かな植生に出逢ったことはなかった」

屋久島の山は、数々の経験を積んできたプラント・ハンターの予想をはるかに凌駕する豊富な植生と生命の謎に満ちた森を形成していた。

その森で外国人植物学者と過ごした８日間のすべてを、父は息子に語っていた。島に残ってこの素晴らしい森を守れと諭されたこと、山中に設置した五右衛門風呂をことのほか気に入ってくれたこと、準備した野趣豊かな島の味覚を喜んで食したこと、など。それは長く市助の胸にしまわれていた大切な家族の“物語”であった。

次郎助は、漁や山仕事など島の自然に関わりながら一生を終えた。そして、貧しい家計のなかから苦労して学費を捻出し、息子を学問の道に進ませることでウィルソンとの約束を果たそうとした。息子は教職に就き、県内各地の小学校で教鞭をとり、やがて屋久島の教育長になった。退職するまでの40年間、数えきれないほど多くの教え子たちに森の自然の大切さを伝え続けた。

ウィルソンの足跡をたどっていると、信じられない奇跡のような出来事に遭遇することが度々ある。2013年秋、24歳の父親と82歳の息子が100年ぶりの再会を果たしたその一コマは、思い出すたびに胸が熱くなる、まさにクライマックスとも言える瞬間だった。

「ウィルソン株」の左上に立つ次郎助
Jirosuke on the top of 'Wilson Stump.'

父：牧次郎助　明治23年1月3日～昭和30年1月31日（当時24歳）
Father：Jirosuke Maki Jan.3, 1890–Jan.31, 1955 (24 years old at that time)

A Miracle after 100 Years

"My dear father...," tears welled in his eyes. Ichisuke Maki was at a loss for words the moment he laid eyes on the monochromatic photograph of the 'Wilson Stump.'

A photo exhibition of "Wilson's Yakushima" was held, with my hope of acquiring information by which I could identify the people in the photograph. On the evening of the last day, persuaded by his wife, Ichisuke visited the exhibition hall, and for the first time, he clarified his connection to Wilson.

To show the size of trees, Wilson almost always arranged a person to stand near them when taking photographs. The stump of Yaku–cedar found at the altitude of 1030m on February 22 deep in the mountains had a such remarkable circumference of 15.2m that one person was placed at each of the four corners for the photograph. Aside from the man wearing a uniform, who seems to be a staff member of the district forestry office, I conjectured that the other three men were local guides because of the padded kimono, gaiters, and traditional shoes they were wearing. Ichisuke pointed at the upper left to the man with a boyish–looking face. His name was Jirosuke Maki, the father of Ichisuke. He appeared in most of the photographs taken in Yakushima, and it was evident that he was a young man Wilson felt close to.

Wilson wrote, "Nowhere else have I seen such a wealth of vegetation."

The mountains in Yakushima hold forests that are full of mystery with abundant vegetation and life that even surpasses the expectations of a plant hunter who has acquired considerable experience.

Ichisuke's father had told his son everything about the 8 days he spent in the forest with the foreign botanist, how he had been advised to stay and protect this remarkable forest, how Wilson had enjoyed such things as an iron cauldron bathtub carried into the mountains, and how Wilson had eaten with pleasure the delicacies prepared on this island. It was a precious 'family story' that had been put away in Ichisuke's heart for a long time.

Jirosuke spent his entire life on the island surrounded by Yakushima's nature while he worked as a fisherman and labored in the mountains. Though living on a shoestring, he managed to raise school expenses for his son, and by giving him an education he tried to fulfill a promise that was made to Wilson. His son went into the teaching profession, taught at an elementary school in Kagoshima prefecture, and eventually became the Superintendent of Education in Yakushima. For 40 years until he retired, he continued teaching countless numbers of students the importance of the forests.

As I traced Wilson's footsteps, I often encountered unbelievable, miraculous events. My heart glows whenever I recall the moment the 82–year–old son had met the 24–year–old father for the first time after 100 years in the fall of 2013. It was certainly a highlight of my journey in the footsteps of Wilson.

息子：牧市助　昭和6年2月9日生まれ（現85歳）
Son : Ichisuke Maki, born Feb. 9, 1931 (Current age 85 years old)
写真協力：久保和義　Photo courtesy of Kazuyoshi Kubo.

代々、牧家が住む楠川集落。ウィルソンはこの集落の歩道から山に入った
Kusugawa village where the Maki family have resided for generations. Wilson went into the mountains from the trail entrance at this village.

ウィルソン株　Wilson Stump　*Cryptomeria japonica* D.Don.　1914-02-22

幹周 15.2m のヤクスギの切り株の上に樹高 24m 幹周 3-3.5m の３本のスギが育っている　標高 1030m
Stump of felled Cryptomeria 50ft. round with 3 Cryptomeria Height 80ft. Circumference 11.5&10ft. growing thereon. alt. 1030m

2013-09-23

切り株の中からハート型の空が頭上に浮かぶ
A heart shaped sky from inside of the hollow stump.

発見者であるウィルソンの名が付けられた巨大なヤクスギの切り株。発見当時はコケやシダに覆われ、株の上にはすでに樹齢100年を超えるスギが3本生えていた。伐採時の推定樹齢は3000年、推定樹高は42mと言われる。1586（天正14）年に伐採された時に放置された先端部分は、今でも残っている。

空洞となった中から見上げるとハート型の空が見えることから、多くの人が訪れる人気のスポットとなった。現在は、コケやシダははがされ、株の入り口は来訪者の踏圧や浸食によって根や石がむき出しになっている。

This is the stump of a huge Yaku–cedar that was named after its discoverer Wilson. When it was found, it was covered by moss and various ferns, and already three Cedars, which were more than 100 years old, were growing on top of it. It is said that the estimated 3000 year old tree was 42m high at the time of felling. The tip that was cut off at the time of felling in 1586 still exists on the ground today.

Since one can see a heart shaped sky looking up from inside of the hollow stump, it has become a very popular tourist spot that many visit. Currently, with moss and ferns peeled off, tree roots and stones are exposed at the entrance of the stump due to treading stress by people and erosion.

里山の変遷　消滅した風景

時とともに風景は変わる。それは自明のこととわかっていても、劇的な変化なぞめったに起こりそうもない離島の山さえも大きく変わってしまうのをにわかに理解することは難しい。しかし、山のたたずまいは人の暮らしのありようとともに確実に変わっていく。それが実感できたのは、屋久島の里で撮影されたウィルソンの写真と現在の写真を並べて比較した時だった。

山を下りたウィルソンが宿泊地の宮之浦から北上し、一湊まで歩きながら撮影した里の写真をみると、大正初年の頃の里山はほとんど丸裸ではげ山になっている。ところが、およそ100年を経た現在では山の緑が見事に復活している様子が一目瞭然にわかる。時代の流れに合わせて自然は荒廃するという通説からすると意外な感じがしないでもないが、これは特に屋久島だけではなく日本全国に普遍的に起こった現象だった。

森林、特に里に近い山と日本人は農耕生活を介して長く深い関係を持ち続けてきた。木材は建築、船、道具の材料として利用され、また薪や炭は煮炊き、風呂、暖房、照明のための貴重な燃料、つまりエネルギー源であった。さらに、チガヤは屋根を葺く材料として、落葉や下草や灌木の若芽は牛や馬の飼料や田畑の肥料として欠かせなかった。草地を維持するために、里山への火入れや野焼きも度々行われた。

その結果、水田を取り巻く緩斜面は畑に、なだらかな丘陵地は灌木や草本の原野に、低山帯は柴山や草山になった。そして、このはげ山化は石炭、石油、天然ガスといった化石燃料が使われるようになり、化学肥料が一般に普及するまで続いた。屋久島の場合、1950年代までウィルソンの写真に見られる風景が広がっていた。

やがて里山が放棄されるとともに樹木が侵入し、この半世紀ほどの短い時間で急速に緑を取り戻すことになる。同時に防風林として海岸線を縁取り、人々に落葉や落枝を提供し、草原のなかでひときわ目につく存在であった村のシンボルのクロマツも姿を消した。

今、ウィルソンが歩いた道を辿り撮影地点を検証してみると、山の稜線が写真の背景に写り込んでいなかったら、とうてい特定することは不可能なほど里の風景は変わってしまっているのに驚く。そして、そこに長い間、里山から多くの必要物資を得ることによって生活してきた島人の姿が重なって見える。

センダン　1914-02-26
Melia azedarach Linn.

志戸子　Shitogo　2013-10-10

うず高く積まれた薪の横に立つ樹高9.1mのセンダン。民家の屋根には、強風時に平木が飛ばされないように丸い石がたくさん積まれている。背後の急峻な山に筋状に見えるのは段々畑

Chinaberry (9.1m in height) which stands beside firewood piled in heaps. Many round stones are placed as weight on the roof of houses so that shingles won't be blown off by strong winds. What it looks like striations of land on the mountain side are terraced fields.

A Vanished Scenery

The scenery changes with time. Although this is a common truth, it is not easy to comprehend that mountains as remote as on this island would undergo such drastic changes. Nevertheless, the appearance of mountains will certainly change by the circumstances of people's lives. I realized this when I compared the current photographs with Wilson's photographs of villages in Yakushima.

Examining the photographs Wilson took in 1914 while walking north to Iso from his lodging in Miyanoura, many Satoyama, woodland near villages, were naked and bold. However, it is clear at a glance that vegetation in those Satoyama has revived wonderfully today after 100 years. Although there is a common belief that nature will be ruined as humankind make advances, this was a phenomenon, while not completely unexpected, that not only occurred in Yakushima but all over Japan.

The Japanese have long had a deep relationship with forests around Satoyama through agricultural practices. Wood was used for building houses, boats and various tools. Firewood and charcoal were used as fuel for lighting, cooking, heating rooms, and bathing. They were the only source of energy. Furthermore, Cogongrass was used as material for thatched roofs, and fallen leaves, undergrowth of weeds, and fresh shoots of shrubs were indispensable for fertilizers and feeding cattle and horses. In order to sustain a grassland, open burning was often applied in fields.

As a result, gentle slopes that surrounded irrigated rice fields became fields for growing crops, hillside slopes became prairies for shrubs and herbaceous plants, and low mountain woods were cut for firewood and growing hay. In consequence, the woodland near villages became bare. This bareness of the Satoyama continued until fossil fuels such as coal, oil and natural gas came into use, and as chemical fertilizers became commonly used by the people. In the case of Yakushima, the scenery seen in Wilson's photograph existed until the 1960s.

Before long, Satoyama were abandoned, whereby trees became abundant and vegetation grew rapidly in the short period of half a century. At the same time, the Japanese Black Pine (*Pinus thunbergii*), which occupied the shoreline as windbreak forest and stood out conspicuously on the grassy plains as a symbol of village, had vanished.

As I walked Wilson's path and inspected the photographic sites, I was aware of how it would have been utterly impossible for me to identify the sites without the ridgeline of the mountains clear in the background of his photographs. The image of islanders, who lived there for so long, getting the necessary supplies from Satoyama, overlapped in my mind.

集落のシンボル的存在だった樹高20mのクロマツ。海側に大きなアコウやガジュマルが生え、防風林として使われていた様子が窺える。現在は平地となり、グランドとして使用されている

A Japanese Black Pine (20m in height), which was a symbol of the village. A Banyan Tree (*Ficus superba*) and Chinese Ban-yan (*Ficus microcarpa*) grow at the sea side and it suggests that they were used as a windbreak forest. At present, the field is flattened and used as athletic ground.

クロマツ　1914-02-26
Pinus thunbergii Parl.

志戸子　Shitogo　2013-10-10

宮之浦集落　Village of Miyanoura　1914-02-26

鹿児島県大隅半島から南西60kmの海上にある屋久島は、亜熱帯から冷温帯までの植生が垂直に分布し、「日本の植生の縮図」といわれている。

Yakushima is located in the sea approximately 60km south of Osumi Peninsula in Kagoshima prefecture. Vegetation from the subtropical zone to the cool temperate zone are distributed vertically, and it is said to be the 'Microcosm of Japanese vegetation.'

2013-10-10

川をはさんで宮之浦集落と里山を臨む。里地には段々畑や草地、急峻な里山の山腹以上は薪山があり、山肌は現在と比べ裸にみえる。屋久島の玄関口、宮之浦は河岸段丘に発達した集落で、川沿いはクロマツが植えられ、集落と段々畑の境はアコウやクロマツが防風林となって民家や耕作地を守っていた。当時は、川に橋がかかっておらず船で渡っていた。ウィルソンは川向こうに見える永迫旅館（旧ホテル縄文）に泊まった。

1931（昭和6）年、待望のコンクリート橋が完成。ウィルソンが泊まった総ヤクスギ造りの部屋は、「ウィルソンの間」として長く保存されていたが、2013年夏に老朽化のため解体された。

Facing Miyanoura village and Satoyama woodland across the river, there were terraced fields and grassland at the vicinity of the village, and the steep upper mountainside was utilized for supplying firewood for villagers. The mountain then looks bare compared to the present. Miyanoura, the gateway to Yakushima, developed on the river terrace. Japanese Black Pine was planted along the rivers. Banyan Tree and Japanese Black Pine functioned as a windbreak at the borders of villages and terraced fields, which protected private homes and cultivation land. A bridge had not been built at the time, so boats were used to cross the river. Wilson stayed at Nagasako Inn (currently called Hotel Jyomon).

In 1931, the long–awaited concrete bridge was built. The inn room where Wilson stayed was preserved as the 'Wilson room' until the building was torn down in the summer of 2013.

クロマツ *Pinus Thunbergii* Parl.　**宮之浦**　Miyanoura　**1914-02-26**

海岸から海抜 150m にわたって広がる沿岸草原
The coastal savannah, sea level to 150m.

2013–10–10

現在、深川団地と呼ばれる町営住宅が並ぶ地点から、琴岳を背景に茅場として利用された段丘地を臨む。成長の早いクロマツが生えていた。小河川沿いに薪として利用された常緑樹やアカメガシワなどの先駆性落葉広葉樹の姿も見える。

クロマツは稚樹から樹齢50年程度まで散在していたが、昭和初期に害虫被害、環境の変化などによって個体数が激減した。

A view of the costal terrace with Mt. Kotodake at the background. The costal terrace was once used for cutting grass for roof thatching. Now, town apartments line up. Fast–growing Japanese Black Pine, evergreen tree and deciduous broad–leaved trees such as Japanese Mallotus for firewood used to grow along the small river.

Black Pine trees from saplings to trees 50 years old were once plentiful, but the population has significantly decreased due to harmful insects and environmental change that occurred in the mid–20th century.

志戸子集落　Village of Shitogo　**1914-02-26**

三角錐の丘を背景にした志戸子集落と水田
View of cone–shaped hills with village of Shitogo and rice–fields.

2013-10-10

下り坂の途中から集落を撮影。左下に湿田があり、段々畑が集落の周辺に広がっていた。集落の海側はアコウ、ガジュマル、タブノキからなる防風防潮林に守られている。後背の円錐の山や右側の山は草原や落葉樹の多い低木林で、盛んに柴刈りや草刈りが行われたことを物語っている。電線のように見えるのは電報線。当時、島にはまだ電気は通っていなかった。

現在は集落が拡大した他、山は樹木で覆われて湿田は放棄され、ダンチクやクズが広がっている。左から３番目の円錐の山が消失しているのは、採石のせいと思われる。

The view of the village was photographed from a hill with rice paddy fields down on the left, and terraced fields spread around the village. A wall of Banyan Trees（*Ficus superba*）, Chinese Banyan（*Ficus microcarpa*）and Machilus was planted on the ocean side to protect the village from wind. Cone–shaped mountains in the back and the mountain on the right side consist of prairies and shrub land with many deciduous trees. It seems like mowing and pruning were performed regularly by the villagers.

What appear to be electrical cables were lines used for telegram. Electricity was not supplied on the island in those days.

Today, the village has expanded, mountains are covered with trees, and rice paddy fields are abandoned and filled with Giant Reed and Kudzu（*Pueraria lobata*）. The third cone–shaped mountain from the left has disappeared. It could be the result of a quarry.

オオイタビ *Ficus pumila* Linn. **志戸子** Shitogo **1914-02-26**

志戸子集落近く。鳥居のそばに植えられた３本のヤマザクラの幹をすっぽりと隠す着生植物のオオイタビとフウトウカズラ。標高 30m

Clothing three Cherry trees（*Prunus serrulata*）. Climing Fig（*Ficus pumila*）and Piper kadsura. With torii, near village of Shitogo. alt. 30m

2013-10-10

段々畑の要所につくられた住吉神社。畑の脇にはシュロが植えられているのが見える。オオイタビは海岸沿いの岩のうえにも多数、繁茂していた。現在、神社の鳥居は道の5mほど先に移され、ヤマザクラの姿もなくなっている。

Sumiyoshi Shrine built at a key place in terraced fields. Chusan Palm (*Trachycarpus fortunei*) planted by the fields can be observed. A large number of Climbing Fig (*Ficus pumila*) inhabited the rocks along the shore, too.

Today, the torii of the shrine has been moved 5m from the road and the Cherry tree is gone.

採集日 Collection Date = 1914-02-26

column　「ウィルソン株」の命名者は誰なのか？ Who Named ‘Wilson Stump’?

屋久島を訪れたことのない人でも、「ウィルソン株」の名前は知っている。島の土産物店には、株の中から空を見上げた時に目に入るハート型をパッケージに印刷した多種多様な菓子類が並んでいる。これほど有名になってしまったこの切り株の名付け親は、ウィルソン自身でないとしたら、いったい誰なのか？　多くの人が発する疑問である。

その答えは、当時加治木中学の教員だった植物学者田代善太郎が残した日記にあった。『田代善太郎日記』をひもといてみると、ウィルソンとの接点がはっきりと見える。

1917（大正6）年3月16日（金）「ハーバード大学ウィルソン氏来訪。午前11時48分の列車にてウィルソン氏の来るのを待ち受け…（中略）…自宅に案内して…（中略）…食事ののち、海岸まで散歩し、5時36分帰るを見送る」。実はこの3週間前の2月23日に、田代は鹿児島高等農林学校（現鹿児島大学）でウィルソンに会っていた。ウィルソンの記録では、1917年の2回目の訪日で2月25日から3月12日まで沖縄を探検している。従って、沖縄の旅の前に鹿児島で田代に会い、そこで帰ってきた際に加治木町の自宅を訪問する約束がなされたものと思われる。

6時間ほどの会合で、2人の間にどんな会話がはずんだのか。記録は残っていないが、田代の日記を追っていくと、この日を境に熱心に屋久島の植物資料を集め始めている。そして翌年の8月に初めて屋久島を訪れ、ウィルソンが発見した大株の洞の中で一夜を明かした。1920年には天然記念物調査嘱託の辞令を受け、屋久島調査の希望を提出。教職を辞すと、計4回屋久島を探索して『屋久島の天然記念物調査報告書』を執筆した。「ウィルソン株」の名前が最初に現れる文献である。翌年、屋久杉原生林が天然記念物に指定され、その後の国立公園指定さらには世界自然遺産登録への道筋がこうして整えられた。

屋久島の森の素晴らしさを実感したウィルソンが科学的な調査と保護の必要性を感じ、それを田代に託し、田代はその先鞭に敬意を表して「ウィルソン株」の名前を捧げたと思われる。

Even people who have never visited Yakushima have heard of the ‘Wilson Stump.’ A variety of confectionery packages, with the heart shape view of the sky looking up from inside of the stump, line up at the souvenir shops on this island. If it was not Wilson himself who had named the stump which earned much fame over the years, who on earth was it ? Many people ask this question.

The answer was in the diary of a botanist, Zentaro Tashiro, who was a teacher at a junior high school in Kagoshima at that time. On Friday, March 16, 1917, “Mr. Wilson of Harvard University came for a visit. I awaited his train at 11:48 a.m.... and invited him home.... After a meal, we took a walk to a nearby shore and I saw him off at 5:36 p.m.” Tashiro had met Wilson three weeks before this at the Prefectural Forestry High School, so I presume they made an appointment to visit his home.

There is no record of their conversation, but his diary shows that it was only after this day that Tashiro had eagerly asked for botanical documents of Yakushima. In the following year, Tashiro visited Yakushima for the first time and stayed overnight in the hollow of the large stump of Yaku–cedar which Wilson discovered. He retired from teaching and visited Yakushima four times. Later he wrote “The Natural Treasure Investigation Report of Yakushima.” The name, ‘Wilson Stump’ appeared for the first time in this document. In the following year, Yaku–cedar Virgin Forest was designated a Natural Monument, and eventually was made a National Park. The road to later UNESCO World Natural Heritage Site registration was prepared in this way.

Wilson, who realized the splendor of the forests in Yakushima, felt the need of a scientific investigation and protection, and entrusted Tashiro with the task. Tashiro gave the name, ‘Wilson stump,’ in honor of the inspiration Wilson had given him.

第2章

桜　島

Scene Ⅱ　Sakurajima

桜島　Sakurajima　1914-03-15
桜島噴火による熱風で荒廃したマツ林
Pine woods devastated by hot blast from Sakurajima volcano.

大正初期の鹿児島

屋久島調査を終えたウィルソンが鹿児島に戻ってきたのは２月末日だった。歴史的な大噴火を起こした桜島は今なお、断続的に白い噴煙を吐き出してはいたが、城山の山頂近くに据えられたウィルソンのカメラには、近代都市の景観を整え始めた港町の街並みが写し出されている。当時の鹿児島市の人口は７万人ほど。政治、経済、教育、文化などの社会基盤やインフラ基盤も整い始め、自由で開放感に満ちた庶民のエネルギーにあふれる新しい時代を迎えようとしていた。

1914（大正３）年３月の鹿児島新聞（現南日本新聞）の記事から、ウィルソンが目にした鹿児島の様子を再現してみる。

東京と直結する鹿児島線（肥薩線）が完成し、中央の文化や新しい時代の息吹がこの動脈を通じて入ってくるようになった。また川内、都城、枕崎といった県内の主要地域をつなぐ鉄道網も開通を待つばかりの状態だった。明治末から徐々に整備されてきた国道や県道には馬車、荷馬車が盛んに行き来し、市街地では約1000台の人力車が市民の足として活躍していた。街に電灯が灯り始め、電話の民間加入者も少しずつ増えつつあった。

電気軌道（市電）も街の中心部に向けて延長工事が進められ、地下１階、地上４階の県下初の鉄筋コンクリート造り山形屋デパートメント・ストアの建設がこの頃、着工した。天文館界隈では、劇場「鹿児島座」や常設映画館「メリー座」が新劇や無声映画を催し、新しく出現したビアホール「ロンドン」も勤め帰りの人々で賑わった。

士族男子が多数を占めていた中等教育の門戸も平民や女子に開かれ、中央で起きた護憲運動の影響を受けて鉱山ストライキや農民運動なども起こり、封建的な明治的社会からの脱皮の兆しも見られるようになった。一方、桜島火口に飛び込んだ不倫の男女の顛末が週刊誌並みの興味で連日掲載され、花柳界の芸妓たちのお披露が紹介されるなど大正初期の時代相も垣間見られる。

そんな矢先に起きた桜島大噴火だったが、ウィルソンが訪れた頃には復興への確かな足取りを歩み始めていた。市街の復旧工事も着々と進捗し、市況も回復しつつあった。桜島大爆発記念と称した噴火風景彫刻入りの「懐中時計」、噴火模様入りの「モスリン風呂敷」、そして桜島名物にもなった砂糖菓子「溶岩糖」などの広告が紙面を飾る。また東京、大阪で溶岩を陳列し観覧料を得る人や枯れ木で茶碗や盆などを作っては行商するたくましい薩摩人の姿も見られた。

県内、沖縄、阪神方面のみならず、台湾さらには長崎・門司経由で中国大陸へと航路を延ばす国際港鹿児島も息を吹き返し、岸壁には常時大型汽船が横付けされ、沖仲仕たちの威勢のいい声が湾内に響き渡っていた。

昭和初年の天文館通り（昭和初期の絵葉書）
Tenmon-kan, downtown Kagoshima (picture postcard of the early 20th century).

Kagoshima in 1914

It was the last day of February when Wilson returned to Kagoshima city after his investigation of Yakushima. While Sakurajima was still pumping out white smoke intermittently, Wilson set up his camera on the mountaintop to photograph the port town of Kagoshima, which had begun to take on the appearance of a modern city. At that time, its population was approximately 70,000. Modern social foundations of politics, economy, education, culture and basic infrastructure had begun. Kagoshima was about to greet the new era, which was vibrant with feelings of freedom shared by common people.

From articles in the local newspaper of that time I could form a picture of the Kagoshima Wilson saw in March, 1914.

'The Kagoshima–line,' the railway that connects Tokyo and Kagoshima directly, was completed, and through this artery the central culture and the breath of the new era entered Kagoshima. In addition, a railway network to connect the main areas of the prefecture was about to open. Carriages and wagons came and went busily on streets which had been constructed over the past decade, yet still 1,000 rickshaws played an active part as a means of citizen's transportation. Electric lights began to shine in the town and telephone subscription among private citizens was increasing little by little.

Construction for extending routes of the electric streetcar in the center of the city was pushed forward, and the prefecture's first reinforced concrete building of Yamakataya Department Store started its construction in those days. The theater and a movie house opened in downtown Tenmon–kan, and people on the way home from work stopped by the new beer hall named 'London.'

The doors to the high school which used to be occupied mainly by boys of Samurai families were opened to commoners and girls. Under the influence of Tokyo, there were mine strikes and peasant movements which showed signs of breaking with conventions of feudal Meiji society. On the other hand, the story of an illicit couple who jumped into the crater of Sakurajima was reported like a modern gossip magazine, and names of Geisha in the red light district were listed in the newspaper.

By the time Wilson returned to Kagoshima reconstruction of the city was making steady progress. So were business. Advertisements of 'Pocket Watch' with eruption scenery engraved 'In Memory of the Great Sakurajima Eruption,' muslin cloth with eruption designs, and 'Lava Sugar' (sugar sweets imitating lava) which is a popular souvenir of Sakurajima even now, were in the newspaper. In addition, there were local entrepreneurs who made bowls and trays from burnt trees and others who took lava to Tokyo and Osaka and exhibited them to collect admission fees.

This lively scene included Kagoshima international port which extended routes to Okinawa, Osaka, Taiwan and mainland China. Large steamships were always moored at the quay, and the energetic sounds of the lumper workers echoed around the bay.

1916（大正 5）年 10 月 6 日竣工の山形屋デパート（大正期の絵葉書）
Yamakataya Department Store built in Oct. 6, 1916 (picture postcard of the early 20th century).

1914（大正 3）年 7 月には電気軌道の路面電車が天文館まで延長された（大正期の絵葉書）
The electric streetcar line was laid in downtown in July 1914 (picture postcard of the early 20th century).

鹿児島市　City of Kagoshima　**1914-03-15**

鹿児島市街。遠望に白煙を吐く溶岩と活動中の桜島。右上の黒線はガラス乾板が割れたもの。
Part of city of Kagoshima; smoking lavas and active volcano of Sakurajima in distance.
The line in right corner of photo is a crack in the glass plate negative.

2015-10-20

城山からみた白煙を吐く桜島と鹿児島市街地。大噴火から2カ月経過しても活火山桜島の溶岩には熱があり、水蒸気が上がっている。沿岸部にも白い水蒸気が煙のように立ち昇り、噴火後も活発な様子をみせている。海には蒸気船、帆船も浮かぶ。高い建物がないので、西本願寺鹿児島別院本堂の屋根が目立つ。

1897（明治30）年創建のこの本堂は、1945（昭和20）年6月17日の鹿児島大空襲で焼失。1982（昭和57）年に瓦葺屋根で再建され、4年前に現在のチタン製屋根に葺き替えられた。デザインは昔から変わらない。

A view of part of Kagoshima city from Shiroyama, Sakurajima emitting white smoke in the back. Two months had passed since the large volcanic eruption; Sakurajima was still emitting hot lava and steam. Around coastal areas, the white steam still rises like smoke, looking very actives. A steamship and a sailing boat are floating in the bay. The large tiled roof of Nishi–honganji Temple is conspicuous.

This temple, built in 1897 was destroyed by bombing on June 17, 1945. Rebuilt with a tile roof in 1982, it was finally reroofed with titanium in the same design in 2011.

桜島大正大噴火

鹿児島の街を歩いていると、いつもどこからでも目に入るのがやはり錦江湾（鹿児島湾）に浮かぶ桜島である。時には褐色の山肌に噴煙をたなびかせ、時には青空にくっきりと三角錐の稜線を描き、時には雨霧に身を包むその姿は、間違いなく郷土のシンボルであり、人々の心の拠りどころであり続ける霊峰である。1914（大正3）年1月12日午前10時5分、その桜島が火の山と化した。

噴煙の高さは1万mを超え、火山灰や軽石、溶岩などの噴出物の総量は約20億m^3に上り、その規模も被害も国内では20世紀最大の火山噴火だった。降灰は容赦なく周辺に飛び家屋を埋没させ、軽石は田畑を覆い、火砕流が樹木を焼いた。日本列島を覆った火山灰は、偏西風に乗って遠くカムチャツカ半島でも確認されたという。

夕方にはM7.1の強い地震が発生し、鹿児島市内でも石垣や土蔵が崩れ、道路は崖崩れ、地割れを起こした。一斉に電灯が消え、電話が止まり、鉄道は不通になり、すべての動きが止まってしまった街に灰だけが早春の雨のように音もなく降り注いだ。

「火山は未だ活動的で、溶岩の流出、軽石や灰の量は桁外れだ」

ウィルソンは鹿児島の旅の終盤近く、3月14日に桜島に上陸し、荒廃した風景をカメラに収めた。家屋は焼失し、巨大な溶岩が道を塞ぎ、マツの枯れ木や焼けただれたミカンやビワの木々が累々と並ぶ光景にプラント・ハンターの冷静な目はカメラのフレームを向けた。ウィルソンが撮った4枚の桜島の写真は、自然の驚異の前に身を横たえた植物への無言の鎮魂の祈りが込められているかのようだ。

桜島、市内合わせて死者58名、負傷者112名、倒壊全焼家屋2148戸。これだけの大惨事にも関わらず、比較的人的被害が少なかったのは、火山の変調を察した島人たちが早めに自主避難をしたからであるといわれている。しかし、農地、農作物、住宅といった生活の資産は取り残され、壊滅的な被害を受けた。海を越えることができなかった馬や牛などの家畜も火砕流と灰に呑みこまれた。

溶岩は桜島と大隅半島の間の海峡を埋め、島を陸続きの半島にしてしまったのは同年1月末。連日の噴火と地震が完全に収束したのは、翌年の春だった。

桜島の東方約10kmの牛根村における降灰に埋没した家屋　1914-02-15
Houses buried by ash falls at Ushio village, about 10km east of Sakurajima.

写真提供：鹿児島県立博物館
Photos provided by Kagoshima Prefectural Museum.

地震により亀裂が生じた鹿児島市郊外、甲突川の土手　1914-01-12
A view of a crack on the river bank at the outskirts of Kagoshima, due to the strong earthquake.

The Eruption of Sakurajima

When I walk around Kagoshima, Sakurajima across the bay comes into view almost everywhere. Sometimes smoke hangs over its brown surface, or sometimes its ridgeline stands out in the shape of a truncated pyramid against the blue sky, or sometime veiled by rain and fog; it is definitely the symbol of the city and a sacred mountain in the hearts of people. At 10:05 a.m. on January 12, 1914 Sakurajima erupted cataclysmically. The plume of ash reached more than 10,000m and the volume of ejecta such as ash, pumice, and lava was approximately 2 billion m^3. On account of both the scale and damage, it was the greatest volcanic eruption of the 20th century in Japan. Ash flew to the surrounding area without mercy, houses were buried, pumice covered the farm fields, and pyroclastic flows burned the trees. It is said that the ash which covered the Japanese archipelago was carried by north westerly winds as far as Kamchatka, Russia.

A strong earthquake of magnitude 7.1 was generated in the evening; stone walls and storehouses collapsed, roads were broken, landslides occurred, the earth cracked even in Kagoshima city. The light disappeared all at once, telephones went dead, and the railroad was stopped. Only ash poured into the town where all movement had stopped without a sound.

Wilson wrote "The volcano is still very active. The outpouring of lava, the pumice and ash has been prodigious."

He landed on Sakurajima on March 14 toward the end of his travels around Kagoshima, and photographed the devastated scenery with his camera. Houses were destroyed by fire, huge lava blocked the road; everywhere trees were burnt. Wilson turned his camera to the disaster with a plant hunter's cool eyes, taking 4 photographs, in Sakurajima. Those photographs are like the requiem of a plant hunter for the plants that succumbed to the power of nature.

There were 58 killed and 112 wounded, while 2148 houses destroyed. Human casualties were low because islanders sensed the abnormality of the volcano and evacuated early. However, their assets such as farmland, farm products, and their houses were left behind and suffered crushing damage. The domestic animals such as horses and cows which could not go over the sea were swallowed in pyroclastic flow and ash.

Lava filled the straits between Sakurajima and the Osumi Peninsula by the end of January. Not until spring of the next year was the eruption and earthquake determined to have ended completely.

火山雷を伴った桜島昭和火口の爆発的噴火　**2015-03-29**
An explosive eruption with lightning at Sakurajima Showa crater.

写真提供：成尾英仁
Photo provided by Hideto Naruo.

桜島南東 4.5km の海潟から見た溶岩流　**1914-02-14**
Lave flows descending the southeastern side of the island, seen from Kaigata, 4.5km south-east of Sakurajima.

写真提供：鹿児島県立博物館
Photos provided by Kagoshima Prefectural Museum.

小池集落　Village of Koike　**1914-03-15**

壊滅的な被害を受けた小池集落。背景の桜島との間に溶岩が見える
View of Sakurajima and site of destroyed village of Koike in foreground; lavas in middle distance.

2015-08-04

火砕流によって壊滅した桜島西部の小池集落。発泡した軽石や火山弾などが大量に落下し家屋は破壊され、木々もなぎ倒され、溶岩流が近くを流れた。瓦礫のなかに割れた瓦、つぶれたブリキのバケツ、壊れた戸障子などの残骸が見える。

この頃から、住み慣れた土地を離れて県内のみならず種子島、馬毛島、大隅半島、宮崎県を中心とした日本各地や朝鮮半島に移住を決意した人々を乗せた船が、毎日のように鹿児島の港を後にするようになった。

Village of Koike which was located in the west of Sakurajima was destroyed by pyroclastic flow. Pumice and volcanic bombs were ejected in large quantities. Streams of lava flowed near. Houses were destroyed, trees were snapped off at their bases. The debris of roof tiles, tin buckets, sliding paper screens could be seen.

Ships with people who decided to leave their homeland to immigrate to various other parts of Japan, such as Tanegashima, Mageshima, Osumi Peninsula and Miyazaki Prefecture, and even to Korea, were leaving Kagoshima Port every day around this time.

赤生原　Akoubara　**1914-03-15**

桜島麓。右は高さ 24m の溶岩。荒廃したマツ林が背景の丘に見える
Base of Sakurajima volcano. View of lava 80ft. high and devastated Pine forests.

2015-08-04

右側は押し寄せてきた巨大な溶岩。冷えて固まり一部はちぎれて崩落している。左側の背景の丘の上には、火砕流によって燃えたクロマツ林が見える。緩斜面の段々畑も火砕流で埋められてしまった。噴火以前は山頂火口付近までヤシャブシの林があり、中腹にはクロマツ林や広葉樹の林が広がっていた。

100年が経った今、道路が整備され緑が戻ってきた。御嶽神社の赤い鳥居が道路脇にぽつんと立っている。

On the right side is massive lava which has been solidified by cooling. On the left, the hillside was a Japanese Black Pine forest, burnt by pyroclastic flow. Terraced fields at the foot of the hill had been filled with pyroclastic flow. Before the eruption, there was a forest of Japanese Green Alder (*Alnus firma*) near the summit crater, and the forest of Black Pine and broad leaf trees had spread over the middle mountainside.

100 years later, a road exists and green has returned. Red torii of a shrine is standing alone on the side of the road.

赤生原　Akoubara　**1914-03-15**

桜島の噴火による火砕流で影響を受けた植物。後ろにマツ林、前方にミカンの木と男

Showing effect on vegetation by hot blast from active volcano Sakurajima. Pinus behind; Orange tree with man.

2015-07-29

火山島である桜島の土壌は軽石や火山灰などが空気を含むことで、排水性、保水性、保温性があり、柑橘類やビワなどの果実、根菜類のダイコンの栽培に適している。噴火で桜島島内の多くの農地が被害を受け、これらの農作物は、ほぼ全滅した。

写真は、段々畑に植えられたミカン畑が火砕流に襲われ、荒廃した様子。後背の里山はクロマツ林だったが、同じく枯れてしまっている。

The volcanic soil of Sakurajima since it contains air has characteristics of good drainage, water retentiveness and heat insulation properties. It is suitable for growing fruit such as Orange or Loquats and cultivation of root crops such as Japanese radish. Much farmland of Sakurajima was damaged by the eruption, farm products were almost exterminated.

Photograph shows Oranges planted on terraced fields burned by pyroclastic flow. In the back was a Japanese Black Pine forest, but it was burnt also.

column　大噴火後の植生の復活　Revival of Vegetation

大噴火を起こした桜島では、大量の火山灰と溶岩が同時に2カ所から流出した。また、爆発の初期段階では火砕流が斜面を流れ小池集落などを襲った。

集落の外れの里山は、当時薪採りが盛んに行われ成長の早いクロマツが生えていたが、熱風（火砕流）が吹いたところでは葉はほとんど焼け落ち、枝だけが残った。溶岩が流れた後は表面がごつごつしており、植物体は見当たらない。

溶岩の流出時の温度は1000℃以上あり、植物体をことごとく焼き尽くす。冷えて液体から固体になると、体積は収縮して表面にひびが入り割れ目ができる。そのままでは保水性もなく植物は生えることはできないが、胞子で増える地衣類やコケ類が溶岩の表面に付着し始める。この間も噴火によって空から供給された火山灰がたまり、保水性を回復する。その後飛んできた種子からススキ、イタドリやクロマツなどが命を育む。はじめはススキなどの草が優勢だが、次第にクロマツの林に変わっていく。順調に生育すればクロマツは成長して10m近くになっているところであるが、桜島ではマツクイムシが20年近く前から入り、当初からのマツは枯れてしまった。現在のマツは2世代目以降のもので成長は早い。

一方火砕流で覆われた里山は、マツ以外の木は地下部から芽が吹いたり、鳥や風が運んできた種子が成長したりして、今はタブノキなど照葉樹の森となっている。昭和40年以降、薪採りも行われなくなり、そのまま成長して現在の森に成長したと思われる。

避難していた人々が帰ると、建物の瓦礫や堆積した火山岩、軽石などが取り除かれ、流入した溶岩部を除いて、少しずつ畑が復活し集落が蘇った。

When Sakurajima erupted, large quantities of volcanic ash and lava flowed out from two places at the same time. In addition, pyroclastic flow occurred at the initial stage and attacked villages. In those days, fast growing Japanese Black Pine grew around the villages and villagers often went to collect firewood. In the area where hot wind from the pyroclastic flow blew, most of the needles were burnt, but branches were left. The surface of the places where lava flowed is rugged and scraggly. No plants are seen.

Lava when first expelled from the volcanic vent has a temperature of more than 1,000℃ and burns everything, all plants included. When the molten lava cools and become solid, the volume shrinks, the surface cracks, and the cracks develop into crevices. At this stage, plants cannot grow since lava cannot hold water. However, lichen and moss that proliferate by spores attach to the surface and create bedding ground for plants. While this is happening, the volcanic ash helps the ground restore water retentiveness. After this, seeds of Japanese Pampas Grass, Japanese Knotweed (*Fallopia japonica*) and Black Pine were carried by the wind and germinated. At the beginning, the grasses were dominant, however it later turned into a forest of Pine which would have grown to 10m high by now, but pine insects entered Sakurajima about 20 years ago and the first generation died out. The current Pines are the second generations and grow fast.

On the other hand, at Satoyama which were scotched by pyroclastic flow, some roots sent out shoots and seeds carried by wind and birds germinated and grew into an evergreen broadleaf forest of Machilus over the past 100 years. Collecting firewood has not been carried out since 1965.

When the people who evacuated returned, destroyed buildings and volcanic rock and pumice were removed, and the village revived little by little except where the lava had flowed.

第3章
城　山
Scene Ⅲ　Shiroyama

クスノキ　*Cinnamomum camphora* Nees & Eberm.
城山公園　Shiroyama Park　1914-03-02

都心のオアシス　城山公園

鹿児島市内でのウィルソンの定宿は、朝日通りの商業会議所（現産業会館）隣にあった山城屋旅館だった。この通りはもともと草深い細道だったのを、西南戦争の翌年に、湾に直結する広い道路に作り替えられたものである。海岸線は今よりも内陸にあり、銀行や白壁の倉庫、老舗の問屋などが軒を並べる商港随一の賑やかな界隈だった。

3月2日早朝。灰まじりの細雨が降り注ぐなか、ウィルソンは蝙蝠傘を小脇に抱え、火山を背に朝日通りを西に向かって歩き始めた。目の前に見える小高い丘は、標高103mの城山である。同じ場所に立ってみると、通りの両端に桜島と城山が均衡よく対峙している様子が実感できる。通りはやがて低い石垣と堀が並ぶ磯街道と直角に交わる。現在、石垣の向こうは歴史資料センター黎明館、図書館、美術館、博物館などの公共施設が並ぶ文化ゾーンになっている。

この一帯は、島津家77万石の藩主の居城である鹿児島城だった。城と言っても天守閣のない低い屋形造りで、翼を広げた鶴に姿が似ていることから鶴丸城と呼ばれた。かつての戦時用の陣所（上山城）があった城山の山頂に登ると、桜島を背景に市街と錦江湾が一望でき、この城を造った人の思想がわかる。海からこの山麓の緑に隠れた居城を確認するのは難しかったに違いない。

文字通り城付きの山であった城山は、立ち入りを禁じられた区域であったため豊かな自然林が残されたが、数々の歴史的事件に遭遇し現在に至っている。1874（明治7）年、城が消失する大火が発生し、3年後には西南戦争最後の激戦地となって荒廃がかなり進んだ。県下初の県営公園に指定された後、市に払い下げられたのを契機に、照国神社東口から山頂まで遊歩道が整備されたのは1910（明治43）年のことだった。ウィルソンはこの遊歩道に沿って山を登りながら10枚の写真を撮った。

遊歩道は1930（昭和5）年から車道への拡幅工事が進められたが、「城山を守れ！」の激しい反対運動が起こった。その収拾に文化庁が調査を行い、翌年、城山は天然記念物に指定されて危機一髪のところで救われた。その後、空襲で山の林相はすっかり変化し、戦後は背後に団地が、山頂にホテルが建設されるなど開発が進んだ。さらには台風、降灰、土砂崩れ、火災などに何度も見舞われた。枝が落ち、白い幹だけを残して立つクスノキの老樹が、そのすべてを物語っている。

現在、城山は約600種の植物が生育する自然林に覆われている。ウィルソンの写真に残された樹木を特定することは難しかったが、数々の受難を潜り抜け、かろうじて生命を保ってきた木も何本か見つかった。ゆったりとした傾斜が続く遊歩道を歩くと、街の喧騒が嘘のように遠のく。都市の真ん中にこれほど豊かな植物が茂っているところは、全国にも例がないといわれている。

城山展望台から桜島を望む（昭和初期の絵葉書）
View of Sakurajima from the Shiroyama observatory (picture postcard of the early 20th century).

同位置より撮影した現在写真　2015-11-29
Photographed from the same position.

An Oasis in the City

In Kagoshima, Wilson stayed at Yamashiro-ya Inn located on Asahi Boulevard. This street, once a narrow grassy path was made into a straight road that connected directly with the bay in 1878. The shoreline was then further inland and it was the one of the busiest streets of the mercantile port district where banks, white plastered warehouses, and long-established wholesale dealers lined up.

Early morning on March 2, Wilson set out on Asahi Boulevard carrying an umbrella, heading west with the volcano to his back while light rain mixed with ash fell. The hill in front was Shiroyama of 103m above sea level. Now as I stand in the same place I can see well that Sakurajima and Shiroyama face each other along that straight boulevard. This street crosses at right angle with a busy street bordered by stone walls. Currently, on the other side of the stone wall is a cultural zone containing public facilities such as the Prefecture Historical Materials Center, a library, the Museum of Modern Art and a Science Museum.

This area was once Kagoshima Castle of the Shimadzu clan. This castle was a low structure without a tower and since it resembled a crane with its wings open, it was called Tsurumaru-jyo (crane castle). I can understand the philosophy of the builder when I climb up Shiroyama to where the military post was located. From the hill top, I can overlook the city and Bay of Kagoshima with Sakurajima at the back. It would have been difficult to see this castle from the sea since it is hidden by forest at the foot of the hill.

Shiroyama literally means mountain of the castle and the mountain was a sanctuary where entrance was prohibited; because of that the rich natural forests were left. However, it has suffered a number of historical events. First was a large fire in 1874 which burned down the castle. Three years later it was the site of the last hard fought battle of the Satsuma Rebellion, causing further devastation. In 1910 a promenade was constructed when this place was made a city park. Wilson walked along the promenade to the mountaintop and took 10 photographs.

Plans for expanding this promenade into a motorway were made in 1930, but intense opposition occurred to protect Shiroyama. The Agency for Cultural Affairs finally designated it a Natural Monument in the next year. Thus the promenade was saved. The forest of Shiroyama has suffered from air raid attacks during WWII, changes such as a housing complex at the back and a hotel on the mountaintop. It has been hit by typhoons, ash fall, fire and landslides. The old white branchless Camphor tree standing there can tell you all about it.

The old forest, inhabited by some 600 kinds of plants, still survives. Though it was difficult to identify many of the trees in Wilson's photographs, there are trees that have survived through many hardships, barely keeping alive. When I walk the slightly inclined promenade, the noise of the town fades away incredibly. There is likely no other place in the whole country where such rich plants grow thick in the middle of the city.

城山遊歩道を散策する女性（昭和初期の絵葉書）
Women taking a walk through Shiroyama promenade (picture postcard of the early 20th century).

現在の城山遊歩道　2015-11-29
Photograph of the present promenade.

クスノキ　*Cinnamomum camphora* Nees & Eberm.　**城山公園**　Shiroyama Park　**1914-03-02**

クスノキの古木。樹高 24m 幹周 4.6m
Old camphor trees. Height 80ft. Circumference 15ft.

2015-07-29

城山には森をつくらず孤立しているクスノキの大木が多い。この写真のクスノキも木の下の人と対比してみると、その大きさがよくわかる。このような大径木であれば、水平方向に枝と葉が充満するはずであるが、公園整備で有用な樹種だけを残して下刈りを行っているため、写真のような樹形をとっている。一帯は皇族による記念植樹が多く、手入れ管理がほどこされていて、この木も保護のためか、100 年前にすでに鉄条網で囲われていた。

現在写真のクスノキは同じ木ではないが、生育地点や樹齢などが酷似していたため、参考に撮影した。

In Shiroyama there are many huge Camphor trees standing alone. By comparing the Camphor tree of this photograph with the person under the tree, you'll see how big it was. With such a tall tree like this, the branches and leaves naturally grow horizontally, but as the result of park maintenance of pruning and weeding, the tree looks as is in this photograph. For protection, it had already been fenced with barbed wire 100 years ago.

The Camphor tree of the current photograph is not the same tree, but I photographed it as a reference because the habitation spot and age of the two trees were very similar.

バクチノキ　*Prunus macrophylla* S. & Z.　**城山公園**　Shiroyama Park　**1914-03-02**

灰色と茶色のまだらの樹肌。樹高 10.7m 幹周 1.2m 樹冠 7.6m 標高 50m

Bark gray and brown. Height 35ft. Circumference 4ft. Crown 25ft. through. alt. 50m

2014-09-11

城山の園内歩道の真ん中にまっすぐ生えていたバクチノキ。灰色の樹皮がはがれると赤茶けた滑らかな樹肌が現れる。博打(ばくち)にのめり込み、着ぐるみを剥がされていく様子に似ていることから、この奇妙な名前がついたという。ウィルソンが見た時は、新旧の樹皮がまだらになっていた。

100年の記憶の縁にとどまるかのように、現在も遊歩道の内側に立っている。華奢だった幹は太くなり、根はたくましく盛り上がり、柵の礎石を壊す勢いだ。右からの植物がよく成長して日陰となったために、光を求めて左側に曲がってしまっている。

Bakuchi-no-ki (*Prunus macrophylla*) which grew straight in the middle of the path in Shiroyama. The smooth reddish brown tree skin appears when the gray bark comes off. Its Japanese name is literally means 'gambling tree.' It was so named because it resembles a person who, absorbed in gambling, gradually gets stripped of all his belonging. When Wilson saw this tree, it was spotted by old and new bark.

It stands at the edge of the promenade now as if it is trying to hold on to the memory of 100 years. The slender trunk has become big and the roots vigorously protruding from the ground are about to break the cornerstone of the fence. Since the trees on the right grew big and created shade, it bent to the left for the light.

サンゴジュ　*Viburnum odoratissimum* Ker.　**城山公園**　Shiroyama Park　**1914-03-02**

樹高 7.6m 幹周 0.6m 標高 50m
Height 25ft. Circumference 2ft. alt. 50m

2015-07-29

広くて厚い葉をもつ照葉樹のサンゴジュ。右カーブとなっている遊歩道の先に２本の大きなクスノキが見える。サンゴジュは秋に熟す赤い果実が、サンゴのように美しいところから名づけられたという。厚く水分の多い葉や枝は火災の延焼防止に役立つとされ、防火樹として庭木や生垣に用いられることが多い。

城山も空襲、そして台風、降灰、土砂崩れなどの数々の自然災害に見舞われ、今に至る。新旧の写真を見比べると、サンゴジュは忽然とその姿を消したかのようにも見える。

Sweet Viburnum is an evergreen broadleaf tree and has wide thick leaves. At the point where the promenade curves to the right, two big Camphor trees could also be seen. Sweet Viburnum was given the Japanese name ‘Sango-ju’ which literally means ‘a tree of coral,’ because it bears beautiful coral like red berries in autumn. Since its thick leaf contains a large amount of moisture, it was considered to help prevent fires from spreading. It is often used as garden tree and as a hedge planted as a firebreak.

Shiroyama has been hit by air raids and many natural disasters such as typhoons, volcanic ash falls, and landslides. When comparing the old and new photographs, it looks as if the Sweet Viburnum has vanished.

ニガキ *Picrasma quassioides* Bennett. **城山公園** Shiroyama Park **1914-03-02**

樹高 10m 幹周 0.6m　落葉している
Height 35ft. Circumference 2ft. Bare of leaves.

2015-07-29

城山遊歩道で法面が崩落しやすい場所に生えていた落葉樹ニガキは、胃薬として昔から暮らしに身近な存在だった。当時の新聞によると、この日は雨模様で最高気温は13℃と鹿児島の３月にしては肌寒かった。鳥打帽にマント姿で立っているのは、案内役を担った三好哲男。大正時代の男性はなかなかオシャレである。

今はニガキの姿はなく、左側のクスノキの大木の枯れ幹だけが残っている。

Deciduous tree Bitter Wood（*Picrasma quassioides*）grew at the edge of the promenade near a slope prone to collapse. It has been familiar as a stomach medicine since old days. According to the newspaper of that day, it was rainy with the highest temperature of 13 ℃ , chilly for March in Kagoshima. The man wearing a hunting cap with a cloak was Tetsuo Miyoshi who acted as dragoman. Men of those days were quite stylish.

The Bitter Wood does not exist now. The big Camphor tree on the left died and only the trunk remains.

島津斉彬と写真術

島津斉彬（1809–1858）
Nariakira Shimadzu

写真提供：尚古集成館
Photo provided by Shoko Shusei-kan.

危機到来の歴史的局面において、それを乗り越える能力を持った人物が輩出されることが度々ある。開港を求めて各国の船が押しかけた幕末の変動の時代に、日本でも多くの優秀な人物が登場した。なかでも、島津家28代当主島津斉彬は特に傑出した人物であったといえる。科学、哲学、兵学、地理、農政、教育、絵画など、その興味の範囲は限りなく広く、まさに薩摩が生んだ稀有の“ルネッサンス人”といっても過言ではないだろう。

斉彬が藩主となったのは43歳の時で、世子の時代が長い。洋書を読み世界の情勢を研究し、学者と交わってさまざまな分野の勉強を重ねることに時間を費やした。その蓄積が藩主就任後に一気に具体的な事業という形で爆発する。鶴丸城内に製錬所を作って各種の基礎実験を行い、磯の別邸（仙巌園）の集成館工場群で製鉄、紡績、電信、ガラス、陶磁器などを実用化した。造船にも力を注ぎ、図面だけを頼りに日本初の蒸気船も建造した。また、和・欧文金属活字を製作し印刷本の普及にも貢献し、国内外への留学生派遣も図って開国後の原動力となる人材の育成にも励んだ。治世は7年と短いが、斉彬が近代日本の基礎を築いた功績は計り知れない。

城山の西麓にある照国神社は、この斉彬を祀るために1864（元治元）年に創建されたものだ。現在は高さ19.8mの大きな白いコンクリート製の鳥居が参道に聳え立ち、正月の初詣や六月灯には数十万人の人が参拝する庶民の神社になっている。ウィルソンが訪れた時は、鳥居の代わりに大きなクロマツが立っていた。

3月3日午後。ウィルソンは、照国神社の社殿を背中にして、クロマツを被写体に参道側の南泉院馬場に向かって三脚を立てた。写真（P70）の左上、うっすらと見える桜島を背景に、噴火に伴う地震で崩れた旧県立興業館の屋根を葺く大工の姿がある。その下方では、着物姿の子供たちの輪の中で初老の女性が書物を開いて座っている。目を右に転じると、「鉄砲製造」「自転車修繕」の看板を掲げた店があり、前方の空き地には人力車が並んでいる。大正大噴火の余韻を残す社頭の日常を見事に写した写真は、まさに100年前の一瞬を封じ込めた1枚であると言える。

さて、好奇心の強い斉彬は写真にも尋常ならぬ興味を持った。写真機と薬品をオランダに注文するなどして、翻訳本を頼りに自ら研究し試行錯誤を重ねて撮影も行った。

「遊戯玩具のように思う者もあろうが、写真術は父母の姿を百年の後に残す貴重な術なり」と語っていたという。

それから半世紀余り、その斉彬の魂が眠る神社の前で同じように写真の価値を認めた異国人が、街の様子をレンズに収めようとしていた。

島津斉彬撮影の図（『照国公感旧録』より）
Illustration of Nariakira taking photographs from the book titled, “Terukuni-ko Kankyuroku.”

Nariakira and Photography

In critical phases of historical change, one often finds a person who excels above all others in shaping events. At the time of changes, the end of Edo period, when ships of foreign countries thronged pushing to open the country to the world, many excellent people appeared in Japan. Above all, it may be said that Nariaikira Shimadzu, the 28th head of the family that governed Satsuma, today's Kagoshima, was a particularly prominent person. His unbounded interests extended to science, philosophy, military science, geography, agricultural administration, education, and art. He was a true 'Renaissance man.'

When Nariakira became lord of Satsuma, he was 43 years old. He spent long years as heir apparent, reading many foreign books and studying the world situation. This accumulation of knowledge exploded after he became ruler in the shape of concrete activities. He built a laboratory in his castle to perform various fundamental experiments and by building factories at his villa, Sengan-en, brought everything into practical use. There was iron manufacture, spinning, telegraphic communication, glass making, and ceramics. He constructed the first Japanese steamship only from plans. In addition, he manufactured type molds for Japanese and European languages, contributing to the spread of typography. He also developed human resources by arrenging to send youths abroad to learn western ways. Many of those became leaders after Japan opened itself to the world. Although Nariakira's reign was tragically only 7 years, his contribution to building the foundations of modern Japan is immeasurable.

Terukuni Shrine standing at the west foot of Shiroyama was founded in 1864 to enshrine Nariakira. Now with its big white concrete torii 19.8m in height, it soars as a symbol of the town. The shrine has become the place where hundreds of thousands of people visit on New Year's Day, summer festivals, and other religious events. When Wilson visited, there was a big Japanese Black Pine instead of torii.

In the afternoon of March 3, Wilson photographed the Black Pine at the front approach of the shrine. On the left of the photograph（P70）, with Sakurajima slightly viewed in the back, there are carpenters repairing roofing damaged by the earthquake accompanying the Sakurajima eruption. Underneath the building, an elderly lady dressed in kimono is sitting with a group of children with a book open on her lap. On the right, there are shops with signboards saying 'gun manufacturer' or 'bicycle repair,' and rickshaws are lined up in front. By showing ordinary life in front of the shrine with lingering after effects of the great volcanic eruption, the reality of life 100 years ago is captured in this photograph.

Nariakira, a person with strong curiosity, had a tremendous interest in photography, too. Ordering a camera and solvents from the Holland, he self-studied photography by trial and error with the help of a translated manual.

"You might think this is a play toy, however, this is a valuable tool that preserves the image of your parents 100 years from now," he encouraged his vassals who were taking up the challenge of photography. A half century later, a foreigner who acknowledged the value of the photograph the same way as Nariakira, stood in front of Terukuni Shrine carefully setting up the tripod for his camera.

斉彬の娘３人の写真。日本初の現存湿板写真といわれる。撮影者不明
A photograph of three daughters of Nariakira, which is said to be the first wet plate collodion photography in Japan. Photographer unknown.

写真提供：尚古集成館
Photo provided by Shoko Shusei-kan.

クロマツ　*Pinus thunbergii* Parl.　**照国神社**　Terukuni Shrine　**1914-03-03**

照国神社社頭。樹高 23m 幹周 4.6m
Entrance to Terukuni Shrine. Height 75ft. Circumference 15ft.

2015-10-24

照国神社の社殿を背にして、参道の通りに向かって撮影したもの。鉄筋コンクリート造りの大鳥居が通りの中央に建てられたのは、1929（昭和 4）年のこと。ウィルソンが訪れた時は、神社創建時にはすでに立っていたという樹齢数百年のクロマツが神社のシンボルだった。高齢のためか、大風のせいか、あるいは大噴火に伴う地震の影響を受けたのか、傾げた太い枝が添え木で支えられている。

近隣のお年寄りの話によると、この老マツは戦前に害虫被害で倒れた。現在は同じ場所にヤマモモが植えられている。

With my back against the main building of Terukuni Shrine, this is a photograph of Torii at the approach to the shrine. It was in 1929 that the big torii of steel reinforced concrete was built. The Japanese Black Pine, which was said to have already been there when the shrine was founded, was several hundred years old and was the symbol of the shrine when Wilson visited there. The venerable old tree, leaning over from age and wind or the earthquakes related to the great volcanic eruption, was braced by poles supporting the trunk and the huge lower branch.

According to the story of an elderly resident of the neighborhood, this old Pine was damaged by harmful insects and fell down before the war. Red Bayberry（*Morella rubra*）is growing in the same place now.

ソテツ　*Cycas revoluta* Thunb.　**1914-03-02**
旧制第七高等学校造士館校庭　Ground of the Seventh High School ‘Zoshi-kan’

樹高 5.5m　池野氏がソテツ精子発見につながる実験をした雌株　高校の校庭
Tree 18ft. Pistillate plant on which Mr. Ikeno’s experiments led to the discovery of spermatozoids in Cycas.

2014-09-11

明治時代に創建された全国８つの官立高等学校は特別な伝統校とされ、創建順に番号を振られナンバースクールと呼ばれた。７番目にあたった旧制第七高等学校造士館は、島津公爵の寄付によって1901（明治34）年に城山の麓、火災で焼失した城山城跡に設立された。写真のソテツは、その校庭にあった。1896(明治29)年、東京帝国大学(現東京大学)の池野誠一郎(1866-1943）はこのソテツの胚珠を観察して、世界で初めてソテツの雄性の生殖細胞が運動性を持つ精子であることを発見。ドイツの学術誌に発表し、学界に大きな衝撃を与えた。

花壇の仕切りの石は中央公民館前に、先駆的な研究に貢献したソテツは旧県立興業館（旧考古資料館）前に移植され、今なお元気な姿を見ることができる。2008（平成20）年に「世界で初めて精子が発見されたソテツ」として、鹿児島県の天然記念物に指定された。

Eight national high schools were founded in the Meiji era. They were numbered in order of founding and therefore called 'number school.' They were considered to be special schools. In 1901 the Seventh High School, Zoshi-kan, was built at the foot of Shiroyama where once the castle destroyed by fire stood, founded by the contribution of Prince Shimadzu. Cycas of the photograph was on the ground of this high school. In 1896, Seiichiro Ikeno (1866-1943) of Tokyo Imperial University (current Tokyo University) observed the ovule of this Cycas and discovered that a male generative cell of Cycas was a kinetic sperm for the first time in the world. He published this in a German academic journal and gave the world's scientific society a big shock.

The stones of the flower bed were moved in front of the central public hall. The Cycas which contributed to a pioneer study was transplanted in front of the former Prefectural Industrial Museum and they are still there in good condition. It was designated Natural Monument by Kagoshima Prefecture in 2008.

旧県立興業館（旧考古資料館） 2016-01-22
Former Prefectural Industrial Museum (Old Archaeological Museum).

ハマセンダン *Euodia glauca* Miquel.　**1914-03-12**
旧制第七高等学校造士館校庭　Backyard of the Seventh High School 'Zoshi-kan'

樹高 15m 幹周 2.1m

Height 50ft. Circumference 7ft.

2014-01-27

薩摩が誇る名門、七高の校庭裏から校内を撮影。手前の貯水槽は、シラス土壌でろ過された城山からの湧水を水源とするもの。貯水槽の側には城山の常緑樹と対照的に、成長の早いアカメガシワやハゼノキなどの落葉樹も生えていた。

1945（昭和20）年、空襲で校舎が全焼。半世紀の歴史で、9279名の優秀な卒業生を送り出した学舎は新制鹿児島大学に統合された。現在は鹿児島歴史資料センター黎明館（左）が建っている。撮影ポイントは黎明館の駐車場裏。貯水槽があった辺りでは、今も山側土手から間断なく地下水が滴り落ちている。

The Seventh High School, a distinguished school that Satsuma was proud of ; this photo was taken from behind the campus. Water for the water tanks in the foreground was supplied by a spring filtered by volcanic ash soil called Shirasu in Shiroyama. In contrast to Shiroyama's evergreen trees, deciduous trees such as fast-growing Japanese Mallotus or Wax Tree（*Toxicodendron succedaneum*）grew near the water tank.

In 1945, the school buildings burned down during the air raid. The school which turned out 9,279 excellent graduates in the course of half a century was integrated into Kagoshima University under the new system of education. Reimei-kan of Kagoshima Prefecture Historical Materials Center is there now. The photographic site is in back of the parking lot. Groundwater still drips from the mountain side continuously.

column　天然記念物　城山の森　The Natural Monument, Shiroyama Park

城山の遊歩道を登ると、まず気付くのはクスノキの巨木だ。城山には樹齢400年を超える植林されたクスノキがあちこちにある。クスノキはかつて高貴な木として、仏像の素材、神社や仏閣の建築材に使われた。防虫剤や火薬の原料である樟脳も含まれているため、軍事や生活への利用を考えながら植林されたものと考えられる。薩摩藩では17世紀中頃に、樟脳をオランダにも輸出していた。

尾根や斜面には自然林といわれるスダジイの森が発達している。自然林は限られた光を有効に利用するため階層構造をつくる。一番上の高木層は巨木になったスダジイ、アラカシやクスノキなどが、その下には亜高木層、低木層が続く。バリバリノキ、バクチノキ、ショウベンノキなど面白い名前を持つ木々もある。最下層の草本層にはヘゴ、シロヤマシダ、カツモウイノデなどのシダ植物やナギラン、サツマイナモリ、ヤマコンニャクが見られる。また、クスノキやスダジイの幹にはボウランやキバナノセッコクなどのラン科植物などが着生し、絶滅危惧植物も多い。

城山は東アジアを代表する典型的な照葉樹の自然林があり、豊かな森林植物相を成していることでも知られる。シロヤマシダ、シロヤマゼンマイなどの名はこの地で発見されたことによる。また、この森は空襲や大火の時、延焼を食い止め、市民の生命や財産を守ってくれた。特に北側の錦江湾面に見られる枯れた枝は、1952（昭和27）年４月の大火を止めたクスノキの先端部である。樹幹は枯れているが根元はしっかりと息づいている。

城山が発見地のサツマイナモリ
Satsuma-inamori (*Ophiorrhiza japonica*) . First discovered in Shiroyama.

写真提供：寺田仁志
Photo provided by Jinshi Terada.

When you walk up the promenade, a huge Camphor tree is the first thing you notice. There are many planted Camphor trees more than 400 years old all around Shiroyama forest. Since the Camphor tree was considered a noble tree, it was once used as material for statues of Buddha and in the construction of shrines and temples. Since it contains camphor, which is the raw material for insect repellent and explosive powder, they were planted in consideration of use for both military affairs and living. Satsuma Domain even exported camphor to Holland in the middle of the 17th century.

A natural forest of Castanopsis is growing on the ridge of the mountain and slope. In natural forests, plants create a hierarchical structure to utilize the limited light effectively. The top layer is occupied with tall trees, namely, Castanopsis, Ring-cup Oak（*Quercus glauca*）and Camphor trees. Below are subtree layer and bush layer including trees with funny names such as Baribari-no-ki（scratching tree）, Bakuchi-no-ki（gambling tree）and Shoben-no-ki（urine tree）. The lowest layer is the herb layer of Tree Fern, Shiroyama-fern, Ctenitis and Nagifolium, Ophiorrhiza japonica, A.kiusianus. On the trunk of Camphor and the Castanopsis grow epiphytic plants in the orchidaceous family such as Bee orchid（*Luisia teres*）and a Chained dendrobium, many of which are designated as endangered species.

Shiroyama also holds the typical evergreen broad-leaved trees representing East Asia. Names such as Shiroyama-fern, Shiroyama-osmunda were given because they were discovered in this forest. This forest also blocked the spread of fire, protecting citizen's life and property. On the north side in the direction toward the bay stands the withered branches of a Camphor tree which stopped a big fire in April, 1952. The trunk is withered but the tree is still alive.

第4章
蒲　生
Scene Ⅳ　Kamo

カヤ　*Torreya nucifera* S. & Z.
八幡神社　Hachiman Shrine　1914-03-04

麓集落と武家屋敷

薩摩藩は武士団を城下に集結させる代わりに、「人をもって城となす」という考えに基づいて独特の防衛網を作り上げた。領地を100余カ所の外城（郷）に分け、武家を住まわせ統治に当たらせた。外城と言っても城があったわけではない。管轄する地頭が務める地頭仮屋を中心に「麓」と呼ばれる武家集落を作り、軍事行政の拠点としたのだ。

３月４日、ウィルソンは重富と蒲生のふたつの麓集落を訪れた。

重富は分家した島津家の私領で、兵法の陣立ての形（鋒矢形）に人家を配置し、防衛のため鍵状に道が作られていた。明治以降の区画整理とともに地割が次第に曖昧になっていったが、特徴的なくの字に曲がった街道の角にさしかかった時、ウィルソンが撮った風景に突然、出くわした。

「オイの店やけど……」

奥の畑で鍬を手に腰をかがめる人影があった。近寄って100年前の写真を取り出して見せると、91歳の老人は訝るような口調でそう言った。なんとイヌマキの横に写っていた店舗で1975（昭和50）年頃まで魚屋を営んでいたという。戦後、シベリア抑留から九死に一生を得る思いで故郷に帰ってきたら、周辺のたたずまいは嘘のように変わっていた。

「いつまでも過去を見ていたら、いかん」

老人は凝視していた写真からついと目を離すと、寂しげに笑って畑作業に戻って行った。

蒲生では、地頭仮屋の正門であった御仮屋門の数奇な運命を知った。現在、姶良市役所蒲生総合支所の前にある御仮屋門の前には樹齢約400年のイヌマキが立っている。ところが、ウィルソンがこの門を撮った写真には、どういうわけかこの由緒ある木が写っていなかった。

調べていくうち、この100年の間に門が数回にわたり移動していたことがわかった。ウィルソンが訪れた時は、今の場所よりも100mほど北側にあり、中には女子尋常高等小学校の運動場が広がっていた。なるほど、写真（P82）をよく見ると興味津々の様子でカメラの前に集まっている女の子たちの姿がある。その後、昭和期に八幡神社鳥居角の旧町役場前に、続いて神社東の保育園前に移された。そして1999（平成11）年、役場施設の一部撤去に伴って現在地に戻され、ようやくもともと並んでいた旧知のイヌマキとの再会を果たしたというわけだ。

ウィルソンの撮影地点を探す旅は、そのまま麓町の変遷をたどる道程と重なった。自然石を使った玉石垣や緑豊かな生垣に威風堂々とした武家門。今なお江戸時代の士風の香りが色濃く残る武家集落には、幾世代も前の記憶が影絵のように揺らめいていた。

武家集落のたたずまい。古い武家門や石垣が今も見られる（蒲生）　2014-10-03

The appearance of samurai communities. The old gates of samurai homes and stone walls are still seen. (Kamo)

Samurai Villages

Instead of housing the samurai forces around the main castle, the Satsuma feudal clan developed a unique defense network based on the idea 'people are castles.' They divided the territory into over 100 areas called 'tojyo' which were defensive positions outside of the main castle, and had samurai live and rule those estates. Tojyo literally means 'outside castle,' however there were no castles in them. They built samurai villages called 'Fumoto' around the headquarters of military administration.

On March 4, Wilson visited two of the Fumoto villages, Shigetomi and Kamo.

In Shigetomi, a private fief of one of the Shimadzu's branch families, houses were arranged according to the military strategy of spearhead shaped lines with irregular, angled streets. After the land readjustment of Meiji, much of this arrangement has changed, but when I was approaching a characteristic dogleg street corner, I suddenly encountered a scene that Wilson had recorded in his camera. There was a small shop building next to a line of trees.

In the back field there was the shape of a person stooping down with a hoe in his hands. When I showed him Wilson's photograph, the 91 years old man said, "This was my shop," in a wondering tone of voice. He said he ran a fish shop by the trees until about 1975. He returned home after the war, having barely escaping death in a Siberian detention camp, to find the neighborhood drastically changed. When the old man turned his gaze from the photograph, he smiled sadly, saying "You cannot live in the past forever."

In Kamo, I discovered the curious case of Okariya-mon, which was the gate for the local leader's headquarters. An Inumaki (*Podocarpus macrophyllus*) 400 years old now stands next to the Okariya-mon in front of the Town hall. However, this venerable old tree did not appear in Wilson's photograph of this gate.

After checking this mystery thoroughly, I discovered it had been moved from its original position and later moved again several times. When Wilson visited, this gate was situated about 100m north of the present location in front of a girl's school. Indeed, when I studied his photograph (P82) carefully I noticed there are blurred figures of girls gathered before the camera with great curiosity. Later on, the gate was moved next to the Torii in front of the old town office, and later to the east side of the shrine in front of a nursery school. Finally in 1999 it was moved back to the original location after removal of government office facilities, reunited with its old tree friend.

Looking for Wilson's photographic sites became a journey uncovering the history of the Fumoto villages. In those villages where the ambiance of samurai culture of Edo period still remains strong, the memories of generations past linger like a shadow play.

武家屋敷の庭木の中心には、イヌマキが好まれた（重富）
2014-11-08

The *Podocarpus macrophyllus* was popular among samurai villages and planted as a centerpiece of their garden. (Shigetomi)

シキミ *Illicium religiosum* S. & Z. **重富郷** Shigetomi **1914-03-04**

満開の花　樹高 7.6m 幹周 0.6m
In full flower. Height 25ft. Circumference 2ft.

2014-11-06

採集日 Collecting Date = 1914-03-04

武家集落（麓）だった重富郷の街道沿いで、香気を放つ満開の花を咲かせていたシキミ。後ろの商家は雑貨店であったのか、「一式販賣」の張り紙が見える。建物はビワやイヌマキなどの常緑に混じって、ウメなども一緒に植えられていた。

戦後、区画整理事業などもあって地割り境界も変わり、家老屋敷のたたずまいも少なくなった。シキミの右に立っていた写真のイヌマキは100年の歳月を重ね巨木となり、今も旧士族の庭に立っている。5、6年前に大きく繁茂していた枝が刈り取られた。

The Japanese Star Anise (*Illicium religiosum*) blossoms perfumed the air alongside the street of Shigetomi where was once a samurai village. The house behind might have been a merchant's family. There is a poster that says'for sale'on the sliding door. Ume (*Prunus mume*) trees together with the evergreens such as Loquats and Inumaki (*Podocarpus macrophyllus*) were planted.

Land readjustment after the war changed the allotment of the grounds, and the tasteful appearance of the chief retainer's residence is largely gone. The Inumaki standing on the right side of the Japanese Star Anise grew big over the past 100 years, but the exuberant branches were pruned back five or six years ago.

ソヨゴ *Ilex pedunculosa* Miquel. （**クロガネモチ** *Ilex rotunda*）
蒲生町御仮屋門 kamo Okariya-mon 1914-03-04

村の小学校の校庭にて。緋色の実をつけたソヨゴ（クロガネモチ）。樹高 15m 幹周 2.1m
Grounds of village school. With scarlet fruits. Height 50ft. Circumference 7ft.

2014-11-06

当時、女子尋常高等小学校の正門として使用されていた御仮屋門と、校庭で赤い実をつけてそびえるクロガネモチ（*Ilex rotunda*）。ウィルソンはソヨゴ（*Ilex pedunculosa*）と記載している。クロガネモチの左側にはセンダン、イヌマキの木が見える。黒光りする乳鋲が打ち込まれた、観音開きの威風堂々とした御仮屋門は、もともと蒲生郷（麓）を管轄する地頭が務めた地頭仮屋の正門だった。

移転を繰り返し、17年前に旧地頭仮屋敷地（現始良市役所蒲生総合支所）に帰ってきた。現在の写真の門扉前に立つ樹木は、樹齢約400年のイヌマキの大木。

Okariya-mon was used as the front gate of the girl's school at the time of Wilson's visit. Kuroganemochi（*Ilex rotunda*）stood tall by the gate with red fruit. Wilson recorded it as Soyogo（*Ilex pedunculosa*）. The Chinaberry（*Melia azedarach*）and Inumaki（*Podcarpus macrophyllus*）are seen on the left side of the Kuroganemochi. The majestic Okariya-mon with shiny black ornamental nails on a set of folding doors was originally the gate at the local leader's headquarter who had jurisdiction over Kamo.

After being moved several times, it came back to the original site 17 years ago. The tree which stands in front the gate now is an Inumaki approximately 400 years old.

人里に生きる大クス

年輪を重ねた木の威厳ある立ち姿は、見る者に畏敬の念を与えるものだ。蒲生八幡神社の樹齢数百年のスギ並木に囲まれた参道の階段を登りつめた先で、巨大なクスノキがウィルソンを待ち受けていた。幹の下部に木瘤が隆起し、内部には大きな空洞ができていたが、天に向かって四方に広がる枝には生気あふれる春の新葉が繁茂していた。しめ縄が張られ、周囲を木柵で囲まれているところから、人々に敬われ、大切に保護されてきた様子がみてとれる。

この「蒲生の大クス」は1922（大正11）年に国天然記念物に、1952（昭和27）年に国特別天然記念物に指定され、環境省の日本の巨樹ランキングでは今なお堂々一位の座を誇っている巨木中の巨木である。巨木と言うとやはり注目度が高いのが、世界自然遺産地のコア地域に立つ屋久島の「縄文スギ」だ。樹高は蒲生の大クスとほぼ同じだが幹周が少し小さい。だが、はるかに高齢だ。推定樹齢は7200年という説もあり、科学的に検証してみても少なく見積もって4000年は生きているのではないかとも言われている。内部のいちばん古いところのサンプルが存在しないため、正確な年齢は計測できないのだ。

奇しくも、鹿児島県内にこの2体の巨木が存在するのだが、生い立ちはずいぶん異なる。屋久島のスギが山の厳しい自然のなかで自生する孤高の木であるなら、願いを込めて人の手で植えられた蒲生のクスノキは、里人と長い時間を共有しながら生きてきた木であった。

大クスの現在の樹齢は推定1500年。神社が建立された1123（保安4）年にはすでに神木として祀られていたという。クスノキは地元で特に愛されている木で、鹿児島県内には数多くのクスの大木が見られる。弥生時代から人々の生活に関係してきた有用植物でもあり、神木とされ信仰の対象として保護されてきたことも寿命を保つ要因だったのかもしれない。油脂成分が多く腐朽に強く耐久性があり、加工しやすく、造船材、彫刻材、建材、家具材、楽器、食器として利用され、成分の樟脳は薬用や防虫剤として用いられた。薩摩の風土や文化の中に溶け込み、代々、人々の心に生き続けてきた木ともいえる。

八幡神社の境内奥から東側に延びる遊歩道を行くと、戦没者慰霊記念碑が整然と立ち並ぶ一画に出る。関ヶ原の戦役、西南戦争、戊申の役、日清・日露戦争、太平洋戦争と、記念碑は時間軸に過去をなぞる。人間の歴史は、すなわち戦いの歴史であることを思い知らされる光景である。

巨大な生命体と化した老樹には、100年という年月はつかの間の時の移ろいでしかないのかもしれない。蒲生の大クスはそのすべてを受け止めて、還らぬ人の魂を慰め続けてきたのだろうか。

クスノキ　1914-03-04
Cinnamomum camphora Nees and Eberm.
階段下から撮った大クスの全体像
Photograph of the big Camphor tree taken from the bottom of the stairs.

Giant Camphor of Kamo

Majestic presence of old trees always gives us a feeling of awe. Past a Cedar lined approach and up the stairs leading to Hachiman Shrine in Kamo, a giant Camphor tree was waiting for Wilson. Above its huge bulbous and hollow trunk, great bursts of fresh green leaves of spring were growing on branches which expand in every direction towards the sky. A sacred straw rope surrounded it and a wooden fence protected it. It has been respected and protected with great care by the villagers.

This large 'Giant Camphor of Kamo' was designated a National Natural Monument in 1922, and as a National Special Natural Monument in 1952. It holds first place in the giant tree ranking of the Ministry of the Environment. Jomon-cedar, the greatest Cedar of Yakushima, which is more widely known, is located in the World Natural Heritage Area. Its height is about the same as the Camphor of Kamo but its girth is smaller. However, it is much older, thought to be as much as 7,200 year old. Based on scientific verification, it's estimated to be least 4,000 years, but this still not accurate since a sample of the internal core does not exist.

Marvelously, these two giant trees exist in Kagoshima, but their background is very different. Yakushima's Cedar is a wild tree growing in severe mountain conditions, while the Camphor of Kamo, planted by the hands of people, lives in the village sharing its long life with many generations.

The estimated age of this Camphor tree is 1,500 years. It is said that it was already considered a sacred tree when the shrine was erected in 1123. The Camphor is a tree that is loved by people of Kagoshima, and there are many big Camphor trees in the prefecture. The fact that it was a useful plant, living in close relationship with people since the Yayoi period, (300 B.C.-300A.D.), and is enshrined and protected as an object of faith might have helped it to keep its longevity. It contains a lot of resin which provides strong decay resistance, and since it is strong and can be processed easily it has been utilized for shipbuilding, sculpture, construction, furniture, musical instruments, and tableware. Components of Camphor wood have uses in medicine and as an insect repellent. The Camphor tree flourished in Kagoshima's climate and culture, and has lived in people's hearts for generations.

The promenade at the back of Hachiman Shrine leads to the place where War Memorial Monuments line up. Commemorating the war of Sekiga-hara, the Satsuma Rebellion, Boshin War, the Sino-Japanese War, the Russo-Japanese War, and the Pacific War, these monuments trace a long past. I was made aware how human history is actually the history of war.

For the old giant tree 100 years may only be like the change of seasons for us, but it has encompassed all this long history and keeps on comforting the souls of the people who did not return.

同じアングルから撮った現在写真　2014-10-03
Present photograph from the same angle.

クスノキ　*Cinnamomum camphora* Nees and Eberm.　**蒲生八幡神社**　Hachiman Shrine　**1914-03-04**

日本一の大きさと言われる大クスの根元　樹高 30m 幹周 23m
Base of giant Camphor said to be the largest Camphor tree in Japan. Height 100ft. Girth 75ft.

2014-10-03

国の特別天然記念物にも指定されている本樹は、当時から著名な神木だった。「鹿児島近郊の神社の庭で、日本でいちばん大きいと言われるクスノキに出逢った」と、ウィルソンはそうメモ書きすると、境内からと道路の擁壁側からと２枚撮影した。「蒲生の大クス」は、火災や台風の被害に遭いながらも、樹木医の手で大切に管理され、地域の人々によって心を込めて現在に至るまで保護され続けている。里人はこの巨木を通して、豊作や家族の健康を祈ってきた。

This tree which was designated a National Special Natural Monument, was a well-known sacred tree from early days. “I came across a Camphor tree that is told to be the biggest in Japan in the garden of a Shinto shrine in the suburbs of Kagoshima,” Wilson wrote, and took two photographs of the camphor tree, one from the precincts side and the other from the street side.

Having been damaged by fire and typhoons, this ‘Giant Camphor of Kamo’ has been maintained by specialists and preserved carefully by the local people to the present; they have long prayed for good harvests and health of the family at this tree.

センダンとイチョウ　*Melia azedarach* Linn. & *Ginkgo biloba* Linn.　**蒲生八幡神社**　Hachiman Shrine　**1914-03-04**

左手前にセンダン、右手前にイチョウ
Melia azedarach Linn.（In left foreground）and *Ginkgo biloba* Linn.（In right foreground）

2015-09-04

蒲生八幡神社参道のセンダン（左側）とイチョウ（右側）。後背には常緑樹のクスノキ、スギ、モミ、カヤそして落葉樹のムクロジなどが写っている。今は、センダンの巨木は枯れ、イチョウはさらに大きく成長した。

ウィルソンの写真をよく見ると、地元の小学生だろうか、社殿前の石段に着物姿のたくさんの子どもたちが並んで座っている。1985（昭和60）年、台風の被害で社殿は大破し、現在の社殿はその後再建されたもの。

A Chinaberry（*Melia azedarach*）on the left side and a Ginkgo tree on the right side of the approach to Kamo Hachiman Shrine. Evergreen, Camphor, Cedar, Japanese Fir, Japanese Torreya and the deciduous Mukuroji（*sapindus mukorossi*）appear in the background. Now the Chinaberry is gone, and the Ginkgo tree has grown huge.

Local primary school children dressed in kimonos are sitting in line on the stone stairs in front of the main shrine. The main shrine was completely destroyed by typhoon in 1985, and the current one was rebuilt afterwards.

カヤ　*Torreya nucifera* S. & Z.　**蒲生八幡神社**　Hachiman Shrine　**1914-03-04**

幹周 5.5m
Trunk 18ft. in girth.

2014-10-03

カヤは温帯性の常緑の針葉樹で、鹿児島では1000m以上の標高に生える。精霊が宿る樹とされ、しばしば寺社に植栽される。モミに似ているが、樹皮は縦に長く裂ける特徴がある。

ウィルソンが撮影したカヤは現在も健在で、八幡神社の境内で堂々としたたたずまいを見せている。人の足で根を踏まれることが多かったのだろう、積石で囲まれて根の保護がほどこされている。

Japanese Torreya is a tree of temperate climate and it naturally grows at an altitude of more than 1,000m above sea level in Kagoshima. It is the evergreen conifer often planted in the precincts of temple and shrine as the tree where spirits dwell. It resembles a Japanese Fir but has the distinction of the bark splitting in the longitudinal direction.

The Torreya which Wilson photographed is still in good condition and has a dignified appearance on the grounds of Hachiman Shrine. The exposed tree roots must have been stepped on often by people; they are now protected by a low stone wall.

column　武家集落の植物たち　Plants of Samurai Villages

今も古い町割が残る武家屋敷群。100年前には、どんな植物たちがウィルソンを待ち受けていたのだろうか。

かつて郷士が住んだ蒲生郷の旧家を覗くと、蒲生石と呼ばれる溶結凝灰岩の石垣の上には生け垣が続き、屋敷の中にはイヌマキの大木がある。石垣の石には手鑿（てのみ）の跡が残り、その石の上をびっしりと蔓植物のオオイタビが張り付いている。秋ともなればイチジクに似た実が熟れ、子どもたちは先を争って食べた。

生け垣の樹木は針葉樹のイヌマキやラカンマキが多いが、照葉樹のイスノキや海岸植物のマサキやハマヒサカキも使われた。選ばれた木々は剪定すると切り口近くから潜んでいた芽が出て目隠しとなる。海岸や山地の木々の性質を知っての選択である。

屋敷のシンボルであるイヌマキを芯とし、形を作るところが和の庭の心である。赤い実のなることで豊かさを象徴する樹木として、しばしばクロガネモチが植えられ、低木にセンリョウやマンリョウを添えることもあった。

屋敷の中では女の子が生まれるとツゲを植えた。ツゲの成長は遅いが、20年もたって嫁に行く頃には櫛に加工できるほどの大きさになる。庭の片隅には食べ物を盛り包むハランもみられる。丈夫で清潔、抗菌作用を持つ葉である。また、陽の当たらない屋敷の北面にはナンテンが植えられ、北面に配置されたトイレからの腐敗・分解を抑え、消臭とした。ナンテンは日陰では生育がよいが、乾燥した東南面では逆に成長が芳しくない。シキミは仏前に供える植物として植えられた。香りがあり死臭を消すという。また、お墓に供えるためサカキあるいはヒサカキ、マサキも庭を飾った。

このように武士の庭は、美観と暮らしに役立つ植物に囲まれて合理的に整えられていたのだ。

Old samurai houses still remain in the residential area of Kamo. What kind of plants were waiting for Wilson 100 years ago ?

Surrounding the Samurai home is a stone wall made of welded tuff called ‘Kamo stone,’ and there is a big Inumaki in the garden. Traces of hand chisel are left on the stone, and Climbing Fig (*Ficus pumila*) grows densely here and there on its top. When it was autumn, a fruit like a Fig ripened there and children competed to be the first to eat it.

For hedges, many used Inumaki and Rakanmaki (*Podocarpus chinensis*) but the evergreen broad-leafed Isu tree (*Distylium racemosum*) and the shore plant Japanese Spindle (*Euonymus japonicus*) and Eurya (*Eurya emarginata*) were used as well. When those trees were pruned, they put forth dense leaves and provided a blind. This shows good knowledge of the properties of trees from both sea side and the mountains.

The Inumaki was chosen as the symbol of these Japanese style gardens. The tree which symbolized wealth, Kuroganemochi (*Ilex rotunda*) was often planted, and sometimes shrubs such as Senryo (*Sarcandra glabra*) and Manryo (*Ardisia crenata*) were added, whose Japanese names come from Edo time’s currency.

When a girl was born to the house, a Box Tree was planted. It grows slowly, but when the girl was about to marry, it would become big enough to make a comb. In the corner of the garden, the Cast-iron Plant would be growing. It was used to wrap food, having strong, clean antibacterial leaves. In addition, Nanten (Nandina) was planted on the north side of the house to control odors from the toilet. The Japanese Star Anise (*llicium religiosum*) was planted to be offered at the Buddhist altar to deodorize the scent of death. In addition, Japanese Cleyera, Japanese Eurya, and Japanese Spindle were planted to be offered at the grave yard.

In this way, Samurai gardens are meaningfully arranged for both beauty and utility.

第5章

霧　島

Scene V　Kirishima

アカマツ　*Pinus densiflora* S. & Z.
霧島山裾　Lower slope of Mt. Kirishima.　1914-03-08

火の里　神話の国

3月5日。蒲生から加治木駅まで馬車で走り、列車に乗ったウィルソンは霧島の山に向かった。明治末から立て続けに爆発を繰り返している活火山のひとつ、御鉢が桜島大噴火の4日前に噴石を噴き上げて以来、山全体がすっかり落ち着きを失ったように見えた。噴火の余震も度々地面を震わせ、鳴動が褐色の山肌に響き渡っていた。

大小20座を超える火山が連なる霧島の連山は、上昇気流が激しく霧や雨がよく発生する。麓の街道を馬車で進むウィルソンの眼前には、灰が風に乗って粉雪のように舞い、その灰が雨に打たれると固まり、轍(わだち)の跡に車輪がとられた。

「霧島の旅は常に雨と灰に悩まされ、十分な結果は出せなかった」

無念の思いをボストン宛の手紙で語るウィルソンだが、事実1週間の滞在で撮影が可能だったのは、そのうちの4日間のみ。しかし、収穫は少なくなかった。

「イヌマキとクロマツの自生種に加え、カシ、シイなどの巨木を観察することができた」

ウィルソンは霧島の東、宮崎県側から南麓を経て温泉郷のある西霧島、そしてさらに活動中の御鉢火口へと標高300mから1200mまでを歩き回り、全部で21枚の写真を残した。山の中で単体の木を被写体にしたものは特定が困難だ。それでも、霧島山麓に点在する神社仏閣の樹木を捉えた写真が紡ぎだす“物語”に私たちは耳を傾けることができる。

初代天皇の誕生を伝える創建2500年の狭野神社、高天原から地上に降り立ったニニギノミコトを祀る霧島神宮とその前身の古宮址、そして高千穂峰の山頂に突き立つ「天の逆鉾」を背景に煙を上げる火口など。日本の創世神話の舞台には、古代の物語が現実と緩やかにむすばれる不思議な時空の気配が漂っていた。

生き物のように一息ごとに膨れ上がりながら天に達する暗灰色の噴煙、地響きを立てる大地、あふれでる赤い炎と山を切り裂く稲妻に太古の人々は超越した存在を見ずにはいられなかった。この火の山そのものが、修験者が修行する道場であり、幾世代も継承されてきた人々の自然に寄せる信仰の社だった。そして、そこに聳え立つ古木は神の化身であり、守り神であり、また天変地異の生き証人でもあった。

ホオノキ　1914-03-06
Magnolia obovata Thunb.

アカガシ　1914-03-08
Quercus acuta Thunb.

コナラ　1914-03-06
Quercus sarrata Thunb.

Country of Volcanos and Myth

On March 5, Wilson went from Kamo to Kajiki station by horse drawn carriage, and took the train to Kirishima. Mt.Ohachi, one of the active volcanos in that area erupted just 4 days before Sakurajima's major eruption; the whole region seemed to be in turmoil. Aftershocks of the eruption shook the ground, and rumbling echoed on the brown surface of the mountain.

The Kirishima mountain range includes over 20 various sized volcanic mountains, and since the ascending current of air is often intense, fog and rain are frequently generated. Traveling the road at the foot of the mountain in a carriage, Wilson saw ash falling like powdery snow. When rain hits the ash it gets hardened and the ruts made by other wheels were slowing the movement of his carriage.

"The trip to Kirishima was marred by heavy rains and did not yield result." Wilson recited these thoughts of regrets in his letter to Boston. As a matter of fact, during the week of his stay, he was able to take photographs on only 4 days. Nevertheless he found points of interest.

"In the forests there I saw spontaneous trees of Inumaki (*Podocarpus macrophyllus*) and Japanese Black Pine and also very large Quercus and Castanopsis."

Wilson went first to the east of Kirishima, and then traveled southwest to a hot spring village, and finally to the crater of an active volcano. He climbed from 300 to 1200m above sea level, and took 21 photographs in total. It was very difficult to identify the location of trees he photographed in this mountainous area. However, his photographs of the trees of shrines and temples scattered around Kirishima speak about the history and myths of that region.

Sano Shrine, founded 2,500 years ago, commemorates the birth of the first emperor of Japan. Kirishima Shrine enshrines the great grandfather of the first emperor, Ninigi-no-mikoto, who came down to earth from the land of gods to form Japan. The sacred halberd of the gods stands on top of the Mt.Takachiho in the background of smoking crater of Mt.Ohachi. On this stage of the Japanese Creation myth, I experienced the mysterious feeling in the air from which an ancient story arises.

The ancient people couldn't help but believe the existence of some transcendent being there when they saw the dark gray smoke reaching the sky, swelling like a creature with each breath it took, lightning splitting the red flames, and when they heard the mountains rumbling. This mountain of fire itself was the place where many mountain ascetics trained themselves and people handed down these beliefs for generations. The old trees that soared in the mountains were incarnations of the gods and witnesses of the extraordinary natural phenomena.

イチイガシ　1914-03-07
Quercus gilva Bl.

ミヤマシキミ　1914-03-08
Skimmia japonica Thunb.

イイギリ　1914-03-06
Idesia polycarpa Maxim.

スギ *Cryptomeria japonica* D.Don.　**宮崎県高原町 狭野神社**　Sano Shrine　**1914-03-05**

神武天皇の神社のスギ並木　樹高 46m 幹周 3-4.6m

Jinmu Tenno Shrine. Avenue of trees. Height 150ft. Circumference 10–15ft.

2014-10-20

神武天皇を祀る狭野神社の巨木スギ並木。直線距離で約 2km あった。島津家 17 代当主義弘が朝鮮出兵に際し戦勝を祈願し、凱旋の奉斎として慶長年間に植栽させたもの。

昭和期に頻発した台風の影響で一部枯れたものもあるが、補植されて景観が維持され、今も見事なスギの摩天楼が続く。階段の留石は、100 年の踏圧で形状が変わっているが、今なお日本の創世神話を語る社にふさわしい厳粛な導入路を作り出している。

The avenue of huge Cedars of Sano Shrine which enshrines the first Japanese emperor, Jinmu. It was approximately 2km long. The 17th head of the Shimadzu household, Yoshihiro, prayed for victory on the occasion of dispatching troops to Korea, and on their triumphant return in the end of 16th century had them planted as a votive offering.

Some of the Cedar trees died from damage caused by frequent typhoons in the 20th century, however the appearance was maintained by supplemental planting, and the avenue of wonderful Cedar skyscrapers still remains today. The stone steps are worn from 100 years of use, but they create a solemn path appropriate for the approach to the shrine which represents the creation myth of Japan.

ムクノキ　*Aphananthe aspera* Planch.　**宮崎県高原町 狭野神社**　Sano Shrine　**1914-03-05**

霧島山麓。枝にシダ植物が着生している　樹高 27.4m 幹周 5.2m
Base of Kirishima. With ferns growing on branches. Height 90ft. Circumference 17ft.

2014-10-20

狭野神社の社殿近くに立つムクノキ。上部の枝にはノキシノブ、ヒトツバなどのシダ植物が着生している。並木のスギとは異なる場所に立っているが、推定樹齢（現400年）から、ほぼ同時期に植えられたと思われる。

今は神社が整備され、周辺の森林には常緑樹が繁ってきている。このムクノキは環境省の巨樹・巨木調査では現在、樹高30m、幹周7.8mとなっている。ウィルソンの計測と比べると、この100年で樹高は2.6m、幹周も2.6mほど大きくなった。

Muku Tree（*Aphananthe aspera*）which stands near the main building of Sano Shrine. Ferns such as Weeping fern（*Lepisorus thunbergianus*）, Hitotsuba（*Ryrrosia lingua*）are growing on its upper branches. It stands in a different location from the Cedar trees of the approach, however, in considering its age（approximately 400 years old）, it must have been planted at the same time.

Today the shrine is well maintained and evergreen trees are growing thick in the surrounding forests. According to the survey of giant trees in Japan done by the Ministry of the Environment, the height of this Muku Tree is 30m and the girth is 7.8m now. In comparison with the measurement of Wilson, the tree grew 2.6m in height and 2.6m in girth during the past 100 years.

アカマツ　*Pinus densiflora* S. & Z.　**高千穂河原**　Takachiho–gawara　**1914–03–06**

鳥居と３本マツ　樹高 23m 幹周 2.1m
With torii. Height 75ft. Circumference 7ft.

2015-05-27

高千穂峰の登山道の起点で、霧島神宮の旧社殿跡である古宮址への参道入り口。樹齢40–60年の大きなアカマツが3本立っていたのは、地元のお年寄りの記憶に鮮やかに残っていた。後ろの小丘には、アカマツがまばらに生える疎林が見える。当時、御鉢では小さな噴火と降灰が続いていた。ウィルソンが移動に使った馬車のものと思われる新しい轍と馬のひづめ跡が地面にくっきりと写っている。

現在、一帯はビジターセンターや駐車場が整備され、シンボル的な3本マツは姿を消した。鳥居背後の小丘には、アカマツの林が繁茂している。

This is the entrance to Mt.Takachiho trail and the approach to the ruins of the old shrine building of Kirishima Shrine. The three tall Japanese Red Pines, 40 to 60 years old in Wilson's photo remain vivid in the memory of elderly locals. In the hill behind, the grove where Japanese Red Pine (*Pinus densiflora*) grow sparsely is seen. Small eruptions of Mt. Ohachi and the fall of ash continued in those days. In Wilson's photo the fresh ruts and hoof-marks of horses are imprinted on the ash covered ground. Perhaps those are of the carriage that he used.

The whole area has been developed into a visitor center and parking lot. The symbolic three Pine trees disappeared. On the hill behind the torii, Japanese Red Pines now grow thick.

写真が語る参道の移り変わり

名所旧跡を訪れても、いわゆる絵葉書的な構図は頑なに避けるウィルソンにしては珍しく、平凡とも思える鳥居を撮った1枚の写真がある。どこから誰が見ても、有名な霧島神宮の神殿前の風景であることは一目瞭然で、当初はこれほど容易に特定できる写真もないと、高をくくっていた。ところが、撮影地点の確認に訪れたところ、思わぬ事実に遭遇した。

鳥居は素材こそ木からコンクリートに変わっているが、形状、高さともに100年前と寸分変わらない。参道脇の樹木のたたずまいには、相応の経年の痕跡はあるものの目を見張るほどの大きな変化は見られない。問題は、鳥居の前にあった。ウィルソンの写真では鳥居前の両側に、土産物などの店舗と思われる建物が建っている。しかし現在は、石の階段が直結していて、とても建造物が立ち並ぶ空間など存在しないのである。つまり、地形と雰囲気がずいぶんと異なっている。

この三の鳥居（当時は二の鳥居）から下って県道沿いの大鳥居から朱塗りの神橋までの間に商業施設が立ち並ぶロータリーがある。いちばんの老舗の旅館「あかまつ荘」を訪ねたところ、主人が大正初期に撮られた当時「濱崎旅館」と呼ばれていた頃の写真を見せてくれた。なんと旅館は三の鳥居のすぐ左横にあり、その屋根のひさしの形がウィルソンの写真に写っているものと完全に一致したのだ。さらにそれを証拠づけるかのように、神宮の倉庫から明治末頃に彫られたという銅版が見つかり、そこにも鳥居の左に旅館、右に土産物店が彫られていた。

霧島神社が「神宮」と呼ばれるようになったのは、官幣大社に昇格した1874（明治7）年のことである。誕生は6世紀にさかのぼり、度々の噴火に見舞われ遷社を余儀なくされていたのを18世紀初頭に島津家が復興させ、出陣などの大事の際は必ず祈願が捧げられる社となった。

もともとは、現在は裏参道になっている西方の亀石坂から社殿に至るルートが参詣の道だった。ところが、1887（明治20）年に洪水で橋が流失し通行止めになったため、それまで生活道路として使われていた社殿正面に直結する道が参拝者に利用されるようになった。国家神道の高まりのなか、1927（昭和2）年から神宮境内外は大規模な改修工事が始まる。「由緒ある境内に風致尊厳を損なう状況は恐れ多し」ということで旅館、土産物屋などは下方に移転させられ、神宮駅につながる道が拡幅され、現在の形が生まれた。つまり、ウィルソンは参道ルートが入れ替わる過渡期に偶然立ち寄り、記憶している人も少なくなった歴史の一コマを写真に撮っていたのだ。

大改修を免れた亀石坂では、火の山に祈りを捧げた古人の素朴な思いを映すかのように、人通りが絶えた石畳の苔の上にマツとスギの大樹が蒼い影を落としていた。

霧島神宮に保存されていた明治後期のものとみられる銅版を反転させたもの。当時の第二鳥居前に広がる門前町の賑わいがよくわかる
This copper plate, presumably from the end of Meiji, has been preserved in the Kirishima Shrine. It shows a flourishing gathering place in front of the torii in those days.

写真協力：霧島神宮
Photo courtesy of Kirishima Shrine.

濱崎旅館（現あかまつ荘）の大正初期の写真　A photograph of Hamazaki Ryokan (current Akamatsu-so) about 100 years ago.

写真提供：あかまつ荘　Photo provided by Akamatsu-so.

The Changing Approach to the Gods

There is one unusual picture by Wilson featuring a large torii, since he avoided taking postcard-like photographs even if he had visited sightseeing spots. At a glance it appeared to be the gate way to the famous Kirishima Shrine. I thought it would be easy to identify. However, when I visited the site to confirm it, I encountered a puzzling situation.

The torii is now concrete, but the shape and the height have not changed from what it was 100 years ago. The appearance of surrounding trees shows little change over those long years. The mystery was in front of the torii. In Wilson's photograph, there were buildings that seemed to be souvenir shops on both sides of the approach to the torii. However, currently stone stairs connect directly and there is no room for any buildings; the topography is very different.

There is a traffic circle down below the torii, where commercial facilities line up. Hoping to solve this mystery, I visited the oldest inn there called 'Akamatsu-so.' The owner showed me a photograph which was taken about 100 years ago when the inn was called 'Hamazaki Ryokan.' And bless my heart ! The inn was located to the immediate left of the torii. The eaves of its roof matched perfectly with those in Wilson's photograph. On further inquiry at the shrine, a copper plate engraved at the end of the Meiji era was found in their warehouse which showed souvenir shops and inns on either side of the torii.

In 1874 the Kirishima Shrine was ranked one of the Major Imperial Shrines. Its birth dates back to the sixth century, but it was forced to move because of repeated volcanic eruptions. In the early 18th century, the Shimadzu restored it and it became the Shinto shrine where prayers were made on serious occasions, such as going into battle.

Originally, the route currently called 'Ura-sando' (Rear approach) from Kameishi Slope to the main shrine was the formal approach for the worshipers. However, this route was closed after a bridge was washed away by a flood in 1887. The current route connecting directly to the shrine, which had been a commercial road, came to be used by the worshipers. In a surge of State Shintoism in 1927, large-scale repair works to improve its grounds and surroundings began. It was considered that the commercial activities which spoiled the dignity of the venerable shrine were dreadful. So the inns and souvenir shops were made to move, the road to the train station was widened, and the current form was born. Wilson visited before the changes from the old approach to the existing approach, and he photographed a scene of its history that few people recollect.

Walking the old approach in Kameishi Slope, which escaped the large scale rebuilding, I found huge Pine and Cedar trees casting their shadows on the moss covered stone pavement, bringing to mind images of ancient people who came there to give prayers to the 'Fire Mountains.'

200余年の間、参詣の主道であった亀石坂。現在は裏参道となり、人影も少ない 2014-12-15
Kameishi Slope which was the main approach to the shrine for over 200 years. The old rear path is seldom used now.

クロマツ *Pinus Thunbergii* Parl. **霧島神宮** Kirishima Shrine **1914-03-07**

石の鳥居と樹高 30m のスギ
Tree 100ft. tall. Stone torii and Cryptomeria.

2014-12-15

霧島神宮社殿前の鳥居と後方のクロマツとスギの並木。左に濱崎旅館の屋根のひさしが写っている。現在の鳥居前は長い石段が続く傾斜となっている。1988 年、国道沿いに大鳥居が建設されると「二の鳥居」から「三の鳥居」へと名称も変わった。

ウィルソンの撮影した左右の並木は、現在も健在だ。写真右奥の神木メアサスギは、南九州一帯のスギの祖先とも言われる。参道の木立が成長したため、今はこのポイントからは見えない。

The torii and the path to the Kirishima Shrine edged with rows of Japanese Black Pine and Cedar. The eaves of the roof of Hamazaki Ryokan are visible to the left of the torii. This area has become a slope with long stone steps. When a new, 'Otorii (big torii)' was built along the highway in 1988, the name of this tori was changed from 'Ni-no-torii (second torii)' to 'San-no-torii (third torii).'

The row of tall trees in Wilson's photograph are still healthy. The enshrined deity of Measa-cedar, local Cedar, which is in the back right of his photograph, is thought to be the ancestor of Cedar trees of the South Kyushu area. Since the trees of the approach have grown thicker, it cannot be seen from this angle anymore.

山麓の混交林　宮崎県都城市吉之元町荒襲　Araso　1914-03-07

火山を望む　モミが目をひく山麓の混交林

View of Volcano. Mixed forests around base in which *Abies firma* S. & Z. is conspicuous.

2014-10-20

右に見える円錐形の山が高千穂峰（1573m）、左の平らな山が御鉢（1408m）。南麓の原野からの景観。国道223号線沿いの宮崎県側、荒襲の砂防ダム辺りと思われる。小丘は里山でモミやアカマツの高木が点在するが、枯れているものも目立つ。

現在、小丘は拡大造林でスギ林に変わり、山腹にも植生が回復して草地や低木林が見えるが、浸食が続き裸地のままのところも多い。

The conical mountain seen on the right is Mt. Takachiho (1,573m), the flat mountain on the left is Mt. Ohachi (1,408m). The landscape is the wilderness of the southern foot; it seems to be the area around the present erosion control dam of Araso on the Miyazaki Prefecture side of National Route 223. The hill was Satoyama of Japanese Fir and Red Pines. It was dotted with high trees, but the dead trees are also conspicuous.

The hill has become a Cedar forest after extensive replanting. The mountainside is now dominated by shrubs. The erosion continues and there are many places which are still bare.

御鉢火口　Smoke in volcanic crater of Ohachi　**1914-03-08**

霧島東、活動中のクレーター
Active crater in Higashi-kirishima.

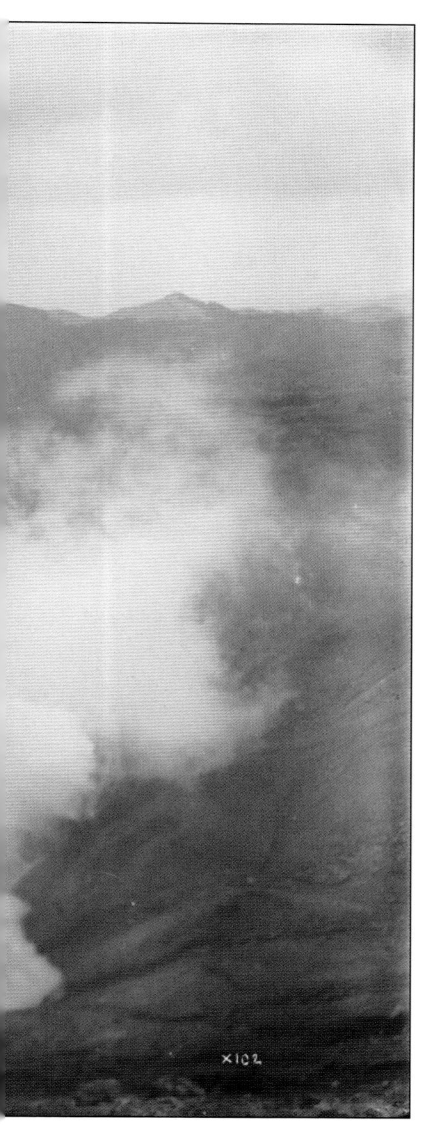

2015-05-27

御鉢は有史以来、噴火を繰り返している活火山で、1923（大正12）年まで火山活動が記録されている。1914（大正3）年1月にも噴火があり、ウィルソンが危険も顧みず撮影を行った日も火口から噴気が立ちのぼっていた。火山活動のため、火口周辺には植物は見られない。

その後火山活動は停止したが、山頂の厳しい環境のため植生の回復は遅い。火口付近は岩の割れ目にスゲやイタドリなどが侵入し、春になればミヤマキリシマが開花し、荒れ地が可憐な赤紫色に染まる。

Mt. Ohachi has been an active volcano since recorded history. Its last volcanic activities were recorded in 1923. It erupted in January of 1914 and when Wilson went there to take photographs, heedless of danger, it was still emitting volcanic gas and steam from the crater. There were no plants around the crater due to the volcanic activities.

Volcanic activity has stopped, however the recovery of vegetation is slow because of the severe environment. In the vicinity of the crater Carex and Japanese Knotweed (*Fallopia japonica*) grow in cracks in the rock, and in spring, the Miyama-kirishima (*Rhododendron kiusianum*) burst into blossoms, dotting the wildness with a pretty violet hue.

ウィルソンが愛したプリンセス

ウィルソンがその花と出逢ったのは、鹿児島の旅を終え、関東まで北上した4月29日、埼玉県戸塚村の園芸店を訪れた時のことだった。中ぶりの鉢に小さくまとめられ枝いっぱいに咲く白い花は気位の高い皇女のごとく、空に向かってりんと首を掲げていた。

「クルメツツジ」と名付けられたその花の稀に見る美しさは、以来、ウィルソンの頭から離れることはなかった。4年後、2度目の来日の際にチャンスが訪れた。台湾の旅から帰るとすぐに、九州は久留米市にある赤司廣樂園まで足を延ばした。1918年5月3日のことであった。

その時の詳細が、ボストンの夕刊紙「イブニング・トランスクリプト」(1920年3月17日)に掲載されている。日本の新聞で言えば4段抜きほどの大きさで添えられているのは、羽織袴で正装した廣樂園主人の赤司喜次郎の写真である。

「満開のクルメツツジを見る心の準備はできていた」とウィルソンは語る。

「しかし繊細な色に染まったその魅惑的な美の世界は、私の予想をはるかに超えていた。欧米の園芸愛好家はこの美をまだ知らないと気づいた時、思わず息が詰まる思いがした」

喜次郎の庭園は、年月をかけて改良された多種多様な色合いのツツジの品種で埋め尽くされていた。ウィルソンは種類、栽培法を詳細に質問し、6枚の写真を撮影した。喜次郎が語ったクルメツツジの来歴にも驚いた。このすべての花の源は、さかのぼること約100年前、坂本元蔵という藩士が霧島の山で見つけたミヤマキリシマを人工栽培して作った品種にあった。坂本の遺志を受け継いだ喜次郎は40年以上にわたってさらなる改良に身を捧げた。まさに日本の庭師の技術と忍耐の記念碑ともいうべき芸術品の数々がそこに生まれた。

ウィルソンは喜次郎を説得して約120鉢の盆栽のクルメツツジを購入すると、細心の注意をはらって梱包させ、ボストンに送った。帰国後、アーノルド樹木園主催の展示会でそのうち50株を披露し、「ウィルソン・フィフティ」と称されるほどの人気を博した。

久留米を出た足で、ウィルソンは霧島山に向かった。クルメツツジの原種であるというミヤマキリシマが咲き誇る姿を一目見たかったからだ。

「同じ色合いの花はふたつとしてなかった。この聖なる山から私のプリンセスが生まれたのです。彼女の最初の恋人であり、庭師である私は、彼女の宇宙における単なる原子にすぎません」

外来種に押されクルメツツジの需要は減り、1873(明治6)年創業の廣樂園は幕を閉じた。ウィルソンのプリンセスはイギリスに移され、今も王立園芸協会の庭園で花を咲かせ続けている。

Kijiro Akashi and His Kurume Azaleas

Wonderful Collection of Japanese Plants Brought by the Famous Collector Wilson from Mount Kirishima and Flowered at the Sargent Greenhouses for the Orchid Show

By George H. Sargent

IT seems almost incredible that nearly three-quarters of a century of intercourse with Japan could elapse before the Occidental nations learned, through the visit of Ernest H. Wilson, assistant director of the Arnold Arboretum, of a beautiful race of garden plants cultivated by this flower-loving nation for generations. Yet at the coming Boston Orchid Show, next week, the visitors will have their first opportunity to behold the rare beauty of the Kurume azaleas, an exhibit of which is to be made by the Arboretum.

There have been a few of these plants in this country for some years, but except to the elect of flower-lovers their existence was unknown. A few were exhibited at the Panama-Pacific Exposition and these were afterwards sold. At Mr. Wilson's suggestion, in 1916, John S. Ames of North Easton secured a number of small plants of these azaleas from Yokohama, and these were the first ever brought into the Eastern States. They have grown well and flowered profusely, and have become one of the floral delights of the Ames estate, a joy to the owner and his friends.

[illegible]

Kijiro Akashi

The Veteran Flower Grower of Mt. Kirishima in Japan, Home of the Kurume Azaleas, Wearing the Decoration Given Him by the Mikado for Flower Culture

[illegible] company and a most important personage in Japanese horticulture. The city of Kurume is on the island of Kyushu and some 800 miles south by west from Tokyo. They arrived there on May [illegible] to find the azaleas in full perfection of bloom.

"I was prepared," says Mr. Wilson, "[illegible] display of blossoms, but the entrancing beauty of myriads of delicately colored flowers covering a multitude of shapely grown plants surpassed my most sanguine expectations. The gardens of Messrs. Akashi and Kuwano were veritable fairylands, and I gasped with astonishment when I realized that garden-lovers of America and Europe knew virtually nothing of this wealth of beauty. Most of the plants were trained into low standards, each about twenty inches high with a flattened or convex crown some eighteen inches through, and were monuments to the patience and cultural skill of the Japanese gardener. Other shapes there were, but this was my favorite and most effective. The flowers, each about half to three-quarters of an inch across, and borne in clusters of from two to several at the end of every twig, were in such profusion as to almost completely hide the leaves. If a fault could be found it was that the flowers were too numerous."

The Garden of Akashi

The focal point in Kurume was the garden of Kijiro Akashi, who for more than forty years has devoted himself to the cultivation and [illegible] of these azaleas

ボストン・イブニング・トランスクリプト新聞
1920年3月17日
Boston Evening Transcript March 17, 1920

Wilson's Dear Princess

On his way north from Kagoshima to the Tokyo region, Wilson visited a nursery in Totsuka village of Saitama Prefecture and there he met the flower for the first time on April 29. The white flowers blooming on the branches, tightly growing in a middle-sized flowerpot, were raising their necks toward the sky like a dignified princess.

Thereafter the singular beauty of the flower named 'Kurume-azalea' had stuck in Wilson's mind. Four years later, visiting Japan for the second time, he went to Akashi's Garden in Kurume City in Kyushu. It was the 3rd of the May in 1918.

Details of that occasion were posted in the daily newspaper, 'The Boston Evening Transcript' on March 17, 1920. A big photograph of Kijiro Akashi, master of the garden, dressed in formal Japanese kimono, appeared in the article.

"I went prepared," said Wilson, "to see a display of blossoms, but the entrancing beauty of myriads of delicately colored flowers covering a multitude of shapely grown plants surpassed my most sanguine expectations. I gasped with astonishment when I realized that garden-lovers of America and Europe know virtually nothing of this wealth of beauty."

The garden of Kijiro was filled with azalea hybrids of various hues. Wilson asked about the plant varieties, cultivation methods in detail, and took six photographs. He was surprised at hearing the origin of Kurume-azalea, Miyama-kirishima, was discovered in Kirishima 100 years before by a clansman, Genzo Sakamoto. He passed the treasure on to Kijiro, who spent over 40 years creating masterpieces that were monuments of Japanese gardening technique.

Wilson persuaded Kijiro to sell approximately 120 pots of Kurume-azalea, packed them with scrupulous attention, and sent them to Boston. After returning home, he exhibited 50 of those at an exhibition hosted by the Arnold Arboretum which won such popularity as to be called the 'Wilson Fifty.'

Wilson headed for Kirishima after visiting Kurume. When he saw a progenitor of Kurume-azalea, in full glory, he did not seem able to control his feelings.

"There were not two flowers of the same hue. My princess was born from this sacred mountain. I, who am her first lover and a gardener, am merely an atom in her space."

The demand for Kurume-azalea decreased as it was pushed aside by the introduction of new foreign species. Akashi's Garden, founded in 1873, closed its doors. However, the princess of Wilson traveled to England and continues blooming in the garden of the Royal Gardening Association.

クルメツツジ　1918-05-03
Kurume-azalea (Rhododendron obtusum f. japonicum Wils.)

ピンク色の花の親株の「吾妻鏡」。樹齢100年といわれている。久留米、赤司廣樂園
'Azuma-kagami' parent of the forms with pink flowers. Plant said to be 100 years old. Mr. Akashi's garden, Kurume.

ミヤマキリシマ　*Rhododendron obtusum* f. japonicum Wils.　1918–05–05

樹高 15–46cm　バラ色の花が溶岩塊の中に咲いている
From 6"–18" high; flowers rose color. Among lava boulders.

2013-06-04

ウィルソンは、1914年の旅ではミヤマキリシマの開花を見ることができず、標本の採集だけ行った。2度目の来日の折に久留米に立ち寄った後、満開のミヤマキリシマを見るために1泊の強行軍で霧島を再訪した。麓で一晩を過ごし、翌日朝早くから出発し、標高1000mの高千穂河原から山に登ったとみられる。山の森は突然、草原に変わり、風が吹く火山岩の傾斜にたくさんの色とりどりの花が点々と咲いていた。

今もこの辺りでは5月から6月の花の季節になると、ローズ色のミヤマキリシマが鮮やかに咲き誇り、カメラを手に訪れる人が絶えない。

On this trip of 1914, Wilson was not able to see the blossoms of the Miyama-kirishima（*Rhododendron kiusianum*）, he only collected specimens of stalk and leaves. However he revisited Kirishima just to see the blossoms on his second visit to Japan. After having stopped at Kurume, he stayed overnight at the foot of the mountain and started to climb early in the next morning from Takachiho-gawara at 1,000m above sea level. The mountain suddenly turned into a grassy plain, and on the windy slope the volcanic rocks were covered with colorful flowers.

Even today, when it is the season for flowers from May to June, the rose-colored Miyama-kirishima provides glorious scenery and there is no end to the throng of people who visit there with camera in hand.

アスナロ　*Thujopsis dolabrata* S. & Z.　**霧島神宮山神社**　Kirishima Shrine　**1918-05-05**

樹高 24m 幹周 4.3m　植栽されたもの
80ft. tall, girth of trunk 14ft. Planted in grounds.

同じアスナロを上から全体を撮ったもの。
右にスギが見える
A picture of the same tree taken from a distance.
Cryptomeria japonica D.Don on right.

2015-11-30

アスナロはヒノキ科の常緑針葉樹で日本の固有種である。青森県ではヒバと呼ばれ、材木として利用されている。鹿児島では一般にあまり見られないが、霧島神宮の境内には、かつてたくさんのアスナロの木が植林されていたという。まっすぐに分岐し、時には縄を編んだかのように交差する独特の縦模様の幹が特徴的だ。

現在は神宮の旧参道沿いにある山の神様を祀る山神社で、祠を守るかのように立つ４本の大木が残っている。ウィルソンが捉えた個体であるという確定はできなかったが、現在写真の木は背景や幹の太さなどから最も可能性のある１本と考える。

Asunaro (*Thujopsis dolabrata*) is an evergreen conifer of the Cypress family, which is indigenous to Japan. It is called Hiba in Aomori prefecture where it is a valuable timber. While not so common in Kagoshima, there were many Asunaro planted in the precincts of Kirishima Shrine. The trunk diverges straight and sometimes has the distinctive feature of resembling a twisted straw rope.

Now four big trees remain at Yama Shrine, which enshrines the mountain god, along the old approach to the Kirishima Shrine as if they protect the small shrine. I could not be sure that this was same tree that Wilson captured in his photograph. However, considering the background and the thickness of the trunk, this was the most likely one.

column　国立公園　霧島の自然　National Forest of Kirishima

霧島山は、1934（昭和9）年に我が国最初の国立公園に指定された。山麓部はシイ、カシなどの鬱蒼とした照葉樹林に包まれ、登るに従ってモミ、ツガ、アカマツの混交林からブナ、ミズナラの落葉広葉樹を主とする森林が続き、その上にはコバノクロヅル、ミヤマキリシマが生育する原野が広がっている。また、霧島温泉郷は約20カ所の温泉地を有し、量、質とも豊かで、美しい森林に包まれた絶好の保養地にもなっている。

数百年前まで溶岩や火砕流流出、大量の火山灰堆積などの火山活動や山崩れがあり、広範囲に裸地ができたところに、まずアカマツが最初の森をつくった。アカマツは樹肌や新芽も赤いことで他のマツと区別ができるが、葉を触ると意外と柔らかくクロマツのように硬くはない。大量の球果がつくが、ひとつの球果から数10個の薄い膜の付いた種子ができる。種子は乾燥すると放出され、風に乗って飛ばされていく。

植物は火山活動直後の無菌的な環境ではなかなか成長しないが、マツは他の樹木に先駆けわずかな時間経過で共生する菌を見つけると急激に成長し、他の樹木を見下す林をつくる。しかし、成長の早いマツも、周辺の樹木が大きくなり受光量が少なくなると衰退し、他の樹木に主役を譲る。

九州では森の主役は標高600mぐらいまでではスダジイやタブノキが多いが、内陸部ではイチイガシが目についた。ところが今はほとんどが開発され、イチイガシが残っているのは人の手が入っていない社寺などの神聖な場所周辺だけになった。

イチイガシは葉の表面は固く、裏面は毛が密生して白い。上に伸びると、成長とともに樹皮が割れ、剥がれては再生を繰り返す。大木になり、気品があるため“一位のカシ”と呼ばれる。材は硬く丈夫でさまざまなものに加工された。また、球果はスダジイに比較してアクがあるためやや渋いが生食もできる。保存性も良いので縄文時代の遺跡からもしばしば発見されている。

Kirishima was designated as the country's first National Park in 1934. The foot of the mountain is covered with dense laurel forests of Castanopsis and Quercus. As we go up, the tree types change to a mixed forest of Japanese Fir, Japanese Hemlock and Red Pines, then to deciduous broad-leaved trees such as Japanese Beech and Mizunara, and at the very top the wilderness where Kobanokuroduru (*Tripterygium doianum*) and Miyama-kirishima dominate. There are about 20 hot springs in this area which are excellent both in quantity and quality, so it has become a popular health resort surrounded by the beautiful forest.

Until several hundred years ago, the volcano was very active and created extensive bare areas where the Red Pine made the first forest. Red Pine is distinguished from other pines by its red bark. The needles are unexpectedly soft and not stiff like those of the Black Pine. It bears a large quantity of cones; one cone produces dozens of seeds with thin skins. When the cones dry, the seeds are released and blown about by the wind. Other plants do not readily grow in the aseptic environment just after volcanic activity, but Pines grow quickly and dominate the forest. However, the fast-growing Pines decline when other trees grow tall enough to catch most of the sunlight, forcing them to hand over their leading role.

In Kyushu, near the coast up to about 600m above sea level, the dominant trees of the forest are Castanopsis and Machilus. Inland, Quercus gilva stood out. However, most of the inland area has been developed, so the Quercus gilva only remain amid the sacred grounds of shrines and temples. The leaves of Quercus gilva are hard with a hairy white underside. Its bark becomes broken and pealed with growth. The wood is hard and strong and was processed into various things. The acorn can be eaten raw. Since it preserves well, it is often discovered in remains of the Jomon period, 14000 B.C–300 B.C.

第6章

川　内

Scene Ⅵ　Sendai

ミツバツツジ　*Rhododendron dilatatum* Miquel.
東郷　Togo　1914-03-14

鎮守の森

ウィルソンは、武駅（現鹿児島中央駅）から川内行きの列車に乗った。川内線の正式な開通はこの年の６月だったが、当時の新聞記事によると３月６日から臨時的に１日１往復、客車と貨車の混合列車が走り始めたとある。薩摩半島を横切って西海岸を北上すると、汽車は約３時間で県下第２の都会であった川内町に着いた。ここで馬車に乗って、１日の行程で菅原神社（藤川天神）と新田神社の２カ所を回った。

川内に限らず、日本の旅ではウィルソンは神社に足繁く通っている。なぜ、神社なのか。特別、日本の信仰文化に興味があったというわけではなさそうだ。プラント・ハンターの目は、神社の周りに広がる「鎮守の森」に注がれていたのである。

古来、日本では森林、山岳や河川、巨石や巨木など自然そのものが信仰の対象とされてきた。つまり神社の社殿や祠にカミが宿っているのではなく、もともとは周りの自然全体がカミの鎮座する聖地であり、人々はそこに社を建て、森を母体として村づくりをした。カミが住む森は、侵してはいけないタブーな聖域であるから、世代を超えて大切に保護されてきた。結果、この鎮守の森には、老樹や原植生と呼ぶべきその地域本来の植生が長い年月、残されるようになった。いわば地域限定型の天然植物園がそこに展開されているともいえる。

ウィルソンは馬車で約４里（16km）、谷と峰を越えて九十九折りの山道を登った先にある東郷の里で、マツとウメが美しい色彩の対比をなす神域、菅原神社を訪れた。そこで、龍が伏したように幹が地上を這っている樹齢1000年の「臥龍梅」と樹高27mに達する巨大なマツに出逢った。参道の両側には、スギやヒノキの大木が立ち並び、社殿の奥はそのまま深い森へとつながっていた。

そして、県内最大の河川である川内川上流では、亀の形に盛り上がった標高70mの神亀山に登った。西方に頭を突き出した亀の左側面、つまり南方から甲羅の上の新田神社まで322段の石段が続く。神社の創建は遙か神代までさかのぼるが、もともと社殿はなく山そのものが神社であったと伝えられる。東麓のニニギノミコトの墓とされる陵（可愛山稜）を中心として、森の各所には田のカミ、山のカミ、穀物のカミなどさまざまな土地の守護神が祀られている。ウィルソンはこの森でクロキ、アケビなど15種ほどの植物を採集した。

1000年、2000年かけて人々が育て残してきた日本各地の鎮守の森には、台風、地震、火事そして戦時中の焼夷弾にも耐えてきた樹木が、今も孤島のような環境のなかで生き続けている。

紫尾山の山懐に抱かれた菅原天神は、鬱蒼とした森に囲まれている。ウメを愛した菅原道真にちなみ、たった１株から繁殖したと言われる約150本の梅林で知られる。２月上旬から咲き始め、多くの梅見客で賑わう

Located in the bosom of Mr. Shibao, Sugawara Shrine, enshrining Michizane Sugawara who loved Ume, is surrounded by the dense forest. This shrine has been well known for the Ume grove of about 150 trees which spread from only one stock. They come into bloom from the beginning of February and are viewed by many visitors.

菅原神社（藤川天神）　Sugawara Shrine (Fujikawa Tenjin)　2014-10-19

Sacred Shrine Forest

On March 14, 1914 Wilson got on the train from Kagoshima city to Sendai. The official opening of the Sendai Line was June of that year, but a train with mixed passenger and freight cars had begun to run one round trip a day from March 6. The train crossed the Satsuma Peninsula and went north along the west coast arriving after 3 hours at Sendai, which was the second biggest city in Kagoshima prefecture. Wilson immediately got on a carriage to Sugawara Shrine (Fujikawa Tenjin) and Nitta Shrine.

In many parts of Japan, shrines were a frequent stop for Wilson. Why shrines ? He did not seem particularly interested in Japanese religion. The eyes of the plant hunter were focused on 'the forest of the local deity' which spread around Shinto shrines.

Since ancient time, nature itself, including the forests, mountains and rivers, huge stones and huge trees, had been objects of reverence in Japan. In other words the gods did not dwell in shrine buildings, rather the natural setting was an enshrined sacred place of the gods. Later the people built shrines there and made villages around the sacred mother forest. Since the forest was where the gods lived and was an inviolable sanctuary they have been protected carefully for generations. As a result, the old trees and original vegetation indigenous to the area came to be preserved for a long time. It might be said that the shrine grounds became the natural botanical garden of the area.

Going by carriage through valleys and around peaks on a winding mountain road for about 16km, Wilson went to Togo village, the site of Sugawara Shrine. The scene was a beautiful contrast of colors where he met a 1000 year-old Ume (*Prunus mume*) that was called 'Resting Dragon Ume' because its trunk crawled on the ground, and also a huge Japanese Black Pine whose height reached 27m. Huge Cedar and Hinoki Cypress (*Chamaecyparis obtusa*) lined the approach to the shrine buildings erected deep in the sacred forests.

Along the upper Sendai River, the prefecture's biggest river, he climbed to Mt. Shingi 70m in height, which was shaped in the form of a tortoise. On the south side of the mountain, 322 stone steps lead to the Nitta shrine standing at the top. The founding of the shrine dates far back to the age of the gods, and there was no building at that time. The mountain itself was the shrine. Grouped around the imperial mausoleum at the mountain top, various local guardian deities such as god of food, god of the mountain, and god of the grain are enshrined and worshiped. Wilson collected about 15 kinds of the plants including Kuroki (*Symplocos lucida*) and Akebia in this forest.

In sacred shrine forests in various parts of Japan, which people established and have preserved for 1000 to 2000 years, the trees which have endured typhoons, earthquakes, fire and wartime bombing are still growing in their environment as if on solitary islands.

新田神社 Nitta Shrine **2015-03-28**

新田神社の神木のクスノキ。樹高約 20m、幹周 9.9m、根回り 13.3m。推定樹齢は 2000 年と言われているが、年輪生長平均率によると約 650–800 年とされる。地上 2m のところに 16 世紀末に掘られた木像が彫刻してある

Camphor tree, the sacred tree of Nitta Shrine. The height is approximately 20m, the girth is 9.9m and the root circumference is 13.3m. The estimated age of the tree is said to be 2000 years old, however, according to dendrochronological analysis, it is about 650 to 800 years old. There is a wooden statue on the trunk 2m above ground which was carved at the end of 16th century.

クロマツ *Pinus thunbergii* Parl.　**菅原神社（藤川天神）** Sugawara Shrine (Fujikawa Tenjin)　**1914-03-14**

樹高 27m 幹周 4.3m　後ろにウメ

Height 90ft. Circumference 14ft. with *Prunus mume* S. & Z. behind.

2015-03-08

国指定天然記念物の臥竜梅の所在地である藤川天神は菅原道真を祀り、正式名を菅原神社という。鳥居の脇には柵によって保護された臥龍梅を背景に、道真の墓と伝えられる塚があり、大きなクロマツが立っていた。後方の小丘は木がまばらに生えている草地だった。

このクロマツは害虫の被害に遭い、1965（昭和40）年に枯死してしまった。現在は残った株に紙垂(しで)を備えた祠が立てられている。古木で樹齢1000年と言われる臥竜梅は、今も3月初旬に淡紅色の八重咲きの花を咲かせ、参道は花見客で賑わう。後方の山は草地から常緑樹の二次林に遷移している。

Fujikawa Tenjin, home of the Natural Monument the 'Resting Dragon Ume,' is dedicated to Michizane Sugawara and its official name is Sugawara Shrine. The Resting Dragon Ume was protected by the fence on the right. Near the torii and what is said to be Michizane's grave stood a huge Japanese Black Pine. The hill at the back was grassland where trees grew sparsely.

This Black Pine died from insect damage in 1965. There is now a small shrine with strips of paper over its sacred remains. The Resting Dragon Ume, said to be 1,000 years old, still puts forth light pink double-petal flowers in early March, crowding the approach to the shrine with flower-viewing visitors. The hill at the back has changed from grasslamd to a secondary forest of evergreen.

ヒノキ　*Chamaecyparis obtusa* S. & Z.　**菅原神社（藤川天神）**　Sugawara Shrine（Fujikawa Tenjin）　**1914-03-14**

中央がヒノキ　樹高30m 幹周2.4m　他はスギ

Height 100ft. Circumference 8ft. in centre. The other trees are Cryptomeria.

2015-03-08

倒木したヒノキやスギで、社殿本体や装飾が造られた
Shrine building and interior ornaments are made of fallen Hinoki and Cedar.

神社社殿のまわりにそびえるヒノキ。参道には、スギおよび落葉しているイチョウが見える。ヒノキは1997（平成9）年の鹿児島県北西部地震で倒木。写真左の夫婦スギと呼ばれていた2本のスギの大木も台風の被害に遭った。命を永らえ、見事な大きさに成長したスギ、イチョウもある。

倒れたヒノキとスギは、拝殿の改修時に建材として活用された。姿を変え、社の一部となって、今も人々の心のなかに生き続けている。

Tall Hinoki Cypresses（*Chamaecyparis obtusa*）rise around the shrine building. Japanese Cedar along the approach to the shrine and Ginkgo that is bare of leaves were seen 100 years ago. The Hinoki Cypresses fell down during the Northwest Kagoshima Earthquake of 1997. Two Japanese Cedars on the left of Wilson's photograph which are called 'Meoto-sugi'（the married Cedar）were damaged by typhoon. Some Cedars and Ginkgo have survived, growing to wonderful sizes.

The Hinoki Cypress and Japanese Cedar that fell down were utilized as building materials for repairing the shrine. In changing form and becoming a part of the shrine, they continue to live in the hearts of people.

ウィルソンとサクラ

ウィルソンが写真に撮ったサクラを見たいと思い、新田神社から開花の知らせが入るのをひたすら待っていた。ウィルソンが訪れたのは3月14日であるから、そのあたりを目安に、まだかまだかと待ちわびていた。ところが、20日を過ぎてもまだ連絡が来ない。私はだんだん気持ちが落ち着かなくなってきた。確かにこのところの地球温暖化に伴う気候変動ははなはだしいものがある。しかし、開花の時期というものは100年の間にそんなに大きくずれるものだろうか。

3月も終わろうかとする頃、ようやく満開の知らせの電話がきた。急いで現場を訪れ、桃色に染まる参道を見た時に、思わず声を失った。そこに誇らしげに咲き誇っていたのは、ソメイヨシノであったのだ。ウィルソンが目にしたのは約半月早く開花するヤマザクラであったから、気候が変わったのではなく、サクラの品種がすっかり変わってしまっていたのだ。

江戸生まれの園芸種であるソメイヨシノは、接ぎ木で簡単に増やせ成長も早いことから、戦後になって爆発的な勢いで全国に広がった。鹿児島県内でも盛んに植栽され、自生のヤマザクラはいつしか花見の主役からはずされてしまったのだ。それでも神社の参道を華やかな色で染めるソメイヨシノの横で、老木の株から育ったヤマザクラの若木が1本、楚々とした風情で色づいている姿を目にした時は、ほっと吐息が洩れた。2015年の春のエピソードである。

欧米で日本のサクラへの関心が高まったのは、20世紀に入ってからと意外と新しい。火をつけたのはアーノルド樹木園の園長サージェント教授で、ウィルソンの日本旅行のミッションのひとつは、さまざまな品種のサクラを集めることにあった。

「サクラは喜びの季節の到来の印である。日本では老いも若きも、金持ちも貧しき人々もサクラの木を愛でる。そして、日本人は100種類以上のサクラを見別ける鋭い目を持っている」

ウィルソンは帰国後、『日本のサクラ』を出版。その中で日本のサクラの多種多様な栽培品種について分類学的に述べている。

日本では約1100年も昔から､原種のサクラを基盤として自然交雑種の収集や人為的交配を行った園芸の歴史があり、現在その品種の数は600以上と言われている。学名の表記や分類も諸説あるが、米国農務省見解のヤマザクラの正名にはウィルソンの名前がつけられている。つまり命名者はウィルソンとされ、最初にその名誉に浴したのが新田神社のヤマザクラであった。

古来、和歌や俳句で詠まれたヤマザクラには、穀物のカミが宿るともいわれ、稲作との関係も深い。赤っぽい若葉とともに春を告げる淡紅色の花弁は、鹿児島の田園風景に良く似合う。

サージェント教授(右)とウィルソン。アーノルド樹木園にて。日本から持ち込まれた満開のエドヒガンの前で
1915-05-02

Ernest Henry Wilson (Left) and Charles Sprague Sargent (right) in front of the full bloomed Edo-higan (*Prunus subhirtella*) brought from Japan at the Arnold Arboretum.

Wilson and Japanese Cherries

I wanted to see the same Cherry blossoms which Wilson photographed, so I waited for news of flowering from Nitta Shrine. It was March 14 when Wilson visited, so targeting that time as the date of my visit I waited impatiently. However, I did not hear anything even by the 20th. I gradually felt unsettled. Climate change resulting from global warming is surely becoming an influence, but how could flowering lag so much after only 100 years ?

When news that the Cherries were in full bloom came on the phone, it was nearly the end of March. I scurried to the site and when I saw the approach to the shrine wrapped in pink, I was at a loss for words. It was Somei-yoshino (*Prunus x yedoensis*) that was blooming there in full glary. What Wilson saw was Yama-zakura (*Prunus serrulata*) that reaches flowering about half month earlier. It was not because the climate had changed but the variety of Cherry had been changed.

Somei-yoshino was a garden variety born in Edo, and since it could be reproduced easily by grafting and grew fast, it spread explosively all over the country after the war. It was planted flourishingly in Kagoshima as well, and the endemic Yama-zakura had been gradually removed from the leading role of flower viewing. I sighed seeing one solitary Yama-zakura, growing from an old trunk, blooming gracefully beside the Somei-yoshino which dyed the approach of the shrine with a gorgeous pink color. It was an episode in the spring in 2015.

Unexpectedly, interest in Japanese Cherry trees grew rapidly in the western world during the early years of the 20th century. It was Professor Sargent, director of the Arnold Arboretum in Boston, who instigated this. He gave Wilson the mission to collect Cherry trees of various kinds during his expedition to Japan. In his 1916 book "The Cherries of Japan" Wilson wrote, "It signifies that spring, the season of gladness, has come. Old and young, rich and poor cherish the Cherry tree. The Japanese have a keen eye for detecting points of difference among their favorite flowers. They recognize more than 100 forms of Cherries." He described many varieties of cultivated Cherry with taxonomic classification.

In Japan there is a history over 1,100 years of the collection of natural hybrids and artificial gardening that crossbred varieties of the progenitor of the Cherry. It is said that the number of varieties is now more than 600. There are various claimants to the scientific name of the Yama-zakura, but the name of Wilson is credited by the United States Department of Agriculture. It was the Yama-zakura of the Nitta shrine that received the honor.

Yama-zakura was often a subject in Japanese poetry since ancient times. It also had a deep relationship with rice-cultivation culture. Its rose pink petals with reddish young leaves foretelling the coming of spring, to me suits well the pastoral scenery of Kagoshima.

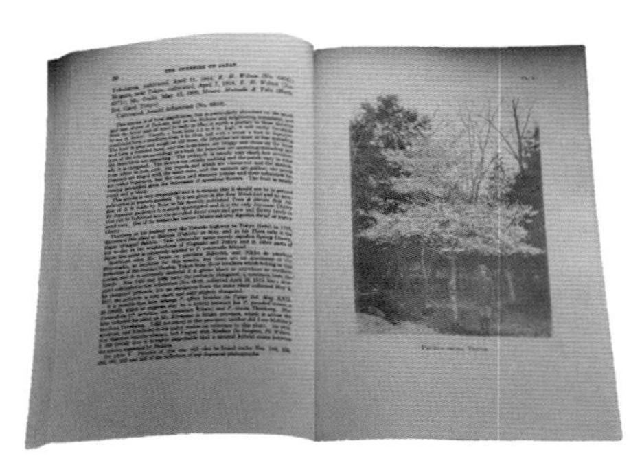

E.H. ウィルソン著『日本のサクラ』 1916-03-30
"The Cherries of Japan" by Ernest henry Wilson, March 30, 1916.

ヤマザクラ　*Prunus serrulata* var. spontanea Wilson.　**新田神社**　Nitta Shrine　**1914-03-14**

新田神社の参道で白い花を咲かせるヤマザクラ　樹高 6–10m 幹周 0.6–1.3m
Entrance to Hachimann Shrine. Flowers white. Height 20-32ft. Circumference 2-4.3ft.

2015-03-28

採集日 Collecting Date = 1914-03-14

標高 70m の神亀山頂の神社に通じる参道の途中まで石段を上りつめた時、突然平地が開け、白く煙るヤマザクラの花霞が目に飛び込んできた。来日の目的のひとつがサクラの研究であったウィルソンは、夢中でシャッターを押し、花びらのついた小枝の標本を作った。

残念ながら、昭和 30 年代にこの平地を横切るかたちで車道が整備され、ヤマザクラの木立は失われた。今はソメイヨシノがほとんどを占めている。後方の山は、オオタニワタリなどが着生した巨木のクスノキが繁る森に変わった。

Climbing the stone stairs to the shrine at the top of 70m high Mt. Singi, Wilson suddenly came to a large level ground filled with hazy white blossoms of Yama-zakura (*Prunus serrulata*). Because one of the objectives was to study Japanese Cherries, he eagerly set up his camera, taking two photographs and collecting specimens of the twigs with the blossoms.

Unfortunately, this grove of Yama-zakura was lost when a road and parking lot was constructed in the 1960s. Somei-yoshino is now the dominant Cherry. The mountain has become a forest containing huge Camphor trees with epiphytic fern such as Spleenwort (*Asplenium antiquum*) growing thick.

column　ソメイヨシノの起源　The Origin of Somei-yoshino

日本の野生のサクラは、ヤマザクラ、オオヤマザクラ、オオシマザクラ、カスミザクラ、エドヒガン、マメザクラ、タカネザクラ、チョウジザクラ、ミヤマザクラの9種で、そこから数えきれないほど多くの品種のサクラが古くから育成されてきた。ソメイヨシノはその代表格であり、ヤマザクラを押しのけて、今や日本の国民的行事である花見の中心的存在となった。気象庁の桜前線、いわゆる3月末から4月初めにかけての開花予想にも使われている。咲き始め、三分咲き、五分咲き、七分咲き、満開、散り始めとその一部始終が刻一刻と報道され、話題となる現象は世界でも珍しい。

ウィルソンが「トウキョウ・チェリー」と呼んだように、ソメイヨシノの生まれは江戸である。名前は、江戸末期から明治初期に染井村の造園師や植木職人たちの手によって人工的な品種改良を経て育成されたことにちなむ。「吉野桜」として売られ広まったが、ヤマザクラとは異なる種であることがわかり、奈良県吉野山に多いヤマザクラと混同される恐れがあり「ソメイヨシノ」と命名されたという。

種子はほとんどできず、挿し木や接ぎ木で増やすクローン植物であるため開花の時期が揃い、葉より先に花が咲いて満開時には花だけが密生して樹木全体を覆う。その派手やかさと、花びらが一斉に落ち桜吹雪が生まれる散り際の見事さが人気を呼び、戦後、他のサクラを圧倒する勢いで多く植樹され、全国に広がった。

このソメイヨシノの起源は長い間不明だったが、ウィルソンは帰国後すぐに発表した論文で、オオシマザクラとエドヒガンの交雑種であるという自説を提唱した。国立遺伝学研究所の竹中要（1903-1966）が交配実験を経て、その仮説の正しさを実証したのは1965年、およそ半世紀後のことだった。

ソメイヨシノ
Cerasus x yedoensis（Matsum.）A.V.Vassil.
写真提供：寺田仁志
Photo provided by Jinshi Terada.

There are 9 kinds of original Japanese wild Cherry trees, which are Yama-zakura, Oyama-zakura, Oshima-zakura, Kasumi-zakura, Edo-higan, Mame-zakura, Takane-zakura, Choji-zakura and Miyama-zakura. The varieties of Cherry that have been bred and cultivated based on them are innumerable. Somei-yoshino（*Prunus x yedoensis*）is now the chief representative among all and plays the central role in the flower viewing events of Japan. It is also used for the Cherry blossom reports that the Japan Metrological Agency announces every spring. The state of flowering is reported almost daily to announce the beginning of blooming, one-third in bloom, half in bloom, three-quarter in bloom, full bloom, and starting to fall. Such a phenomenon is rare in the world.

Wilson called Somei-yoshino the ‘Tokyo Cherry,’ and it was a true child of Edo. The name is associated with having been cultivated after artificial selective breeding around the end of the Edo era by the hands of landscape artisans and gardeners of Somei village. It was sold as ‘Yoshino-zakura,’ however, there were many wild Yama-zakura in Mt. Yoshino in Nara prefecture, and it was renamed ‘Somei-yoshino’ to prevent it from being confused. Since it is a clone plant that is propagated by grafting and bears few seeds, the flowers bloom at the same time. They bloom earlier than the leaf and at the time of full bloom, the flowers are so dense they cover the whole tree. The splendor of its blooming and the rain of its falling blossoms meant its popularity spread all over the country after the war.

The origin of this Somei-yoshino was unknown for a long time, though Wilson had proposed the theory that it was a hybrid of Oshima-zakura and Edo-higan. In 1965, almost a half century later, the National Genetics Institute vindicated this hypothesis after carrying out cross breeding experiments.

第7章

磯

Scene Ⅶ　Iso

ユズ　*Citrus japonica* Thunb.
鹿児島近郊　Near Kagoshima　1914-03-03

近代国家への道のり

鹿児島を後にする前日、ウィルソンは、仙巌園（磯庭園）を訪れた。桜島を築山に、錦江湾を池に見立てた見事な借景の庭園は、鹿児島の旅の締めくくりにふさわしい場所であった。

仙巌園は1658（万治元）年、島津家19代当主光久が別邸として創建し、明治以降は本家の邸宅となった。名前は中国の景勝地、龍虎山の仙岩にちなみ、園内には随所に中国の影響が見られる。光を受けて鏡のように輝く水面に浮かぶ漁船の白帆、そして背景にそそり立つ桜島。なるほど、島津のお殿様が好んだだけあって、ここから眺める湾内の風景は絶景である。1724（享保9）年に山懐を開削して磯街道が作られ、その後鉄道も走るようになったため、高い石垣で海側を囲まれるようになった。昔は、庭が海岸まで続いていたという。5万m^2にもおよぶ広大な敷地には、国内外から移植された多種多様な樹木や花が配置され、温室も整備されていた。さながら植物園のような趣向で、プラント・ハンターの目を楽しませてくれたに違いない。

ウィルソンが足を止めたのは、御殿東奥の石垣の前に立つ1本の大木だった。

「有名な島津のプリンスの庭で、初めてナギを見た」

この木は現在も変わらぬたたずまいで、清水の流れを見下ろす場所に立っている。そして、その足元で毎年春になると、やはり中国起源の「曲水の宴」と呼ばれる歌会が開かれる。水流に運ばれる杯が詠み手の前を通り過ぎる間に短冊に和歌をしたためる伝統の行事である。21代当主吉貴が始めたとされるが、その後断絶。1959（昭和34）年に火山灰や土砂に埋もれていた曲水の庭が発掘され、1990（平成2）年に宴が復活した。

幕末、28代当主斉彬は風流だけに飽き足らず、海へのアプローチの好条件を考えて、庭園の西側の敷地にヨーロッパ式製鉄所やガラス工場などを建設して近代化事業（集成館事業）を起こした。そして、雅な歌会が開かれた水流の源で水車を回し、中国古代の家屋を模した「望嶽楼」に装置したスイッチを押して地雷を爆破させ、石灯籠に配管を通して日本初のガス燈を点火させた。島津家代々の別邸は文化継承の地であると同時に、幕末には先駆的な実験の場ともなった。海を越えて外国と交流してきた土地ならではの進取の気風が、この庭園には流れている。

2015年7月5日、ユネスコ世界文化遺産に登録された「明治日本の産業革命遺産」で、仙巌園は製鉄、鉄鋼や造船に関する工場群「集成館」跡などとともに近代化実験やオランダ人、イギリス人らとの交流の舞台として構成遺産に指定された。現在は石造りの機械工場跡が博物館（尚古集成館）として活用され、島津氏700年の歴史を語る膨大な資料が収蔵されている。

旧集成館（反射炉跡）　2015-11-30
Site of Former Reverberating Furnace
大型の大砲を鋳造するための設備。オランダの書物だけを参考に建設された
The large-cannon casting facility. Built only by referring to Dutch books.

島津斉彬愛用の地球儀　Nariakira's favorite globe

写真提供：尚古集成館
Photo provided by Shoko Shusei-kan.

The Way to a Modern Nation

The day before Wilson departed Kagoshima, he visited Sengan-en, a garden of wonderful design which borrows the view of Sakurajima and Kagoshima Bay. This place deserved to be the completion of Wilson's trip to Kagoshima.

The garden was founded by the 19th head of the Shimadzu household, Mitsuhisa, as his villa in 1658 and became the principal residence of the family after the Meiji era. It was named after Sengan, a scenic spot in China, and the influence of China can be seen everywhere. With the white sail of a fishing boat floating on water which glittered like a mirror reflecting the light, and Sakurajima rising up in the background, the lords of Shimadzu were favored here; the scenery looking out over the bay is superb. After a shore road was made by excavating the foot of mountain in 1724 and later a railway, the sea side was surrounded with a high stone wall. It is said that the garden spread out to the shore in the old days. A great variety of trees and flowers from inside and outside the country were transplanted to the very large site of 50,000m^2, and there was also a greenhouse. Like a botanical garden, it must have pleased the eyes of the plant hunter.

関吉の疎水溝　2015-08-11
Sekiyoshi Sluice Gate of Yoshino Leat
溶鉱炉に空気を送ったり、大砲の砲身をくり抜いたりするために水車動力が駆使された。高低差を考えた精密な土木技術の結晶
Irrigation canal for water wheel power to give air to the furnace and hollow out the gun barrel. The outcome of skillful engineering techniques using the elevation difference.

寺山炭窯跡　2015-08-11
Terayama Charcoal Kiln
反射炉に必要な白炭をつくるために約 5km 離れた森に建造された炭窯
The charcoal kiln was constructed approximately 5 km away from Sengan-en to make hard charcoal used as fuel in the reverberating furnace.

It was the big tree standing in front of a stone wall in the east end of the palace, where Wilson stopped.

"The first *Podocarpus nagi* I saw at Kagoshima grows in the famous garden of Prince Shimadzu."

With unchanged appearance, this tree still stands looking down at the flow of clear water. Every spring, a Chinese styled poetry party, 'Kyokusui-no-en' was held along a meandering brook at the base of this tree. It was a traditional event to write a 31-syllable Japanese poem on a strip of paper in the time it took a Sake cup to float past the composer. This event was started by the 21st head of the household, Yoshitaka, but it was discontinued after the garden was buried under volcanic ash and sediment. It was finally excavated in 1959, and the party revived in 1990.

The 28th head of household, Nariakira, may have appreciated the beauty of this garden, but because of its convenient access to the sea, he built various European style mills and industrial buildings on the western side of the premises. At the head of the stream that fed the brook of the poetry parties he installed a water wheel to power his factories. In the elegant 'Bogaku-ro pavilion' he installed a switching devise to activate explosions in an underground mine. He also installed piping into a traditional stone lantern creating the first Japanese gas light. This location for generations of the Shimadzu family had been a place of culture, but with Nariakira it became a place of a pioneer industrial experiment. As the primary point of contact with foreign countries, the spirit of enterprise flowed through this garden.

In July 5, 2015 Sengan-en and the site of the old factory group were recognized as UNESCO World Cultural Heritage Site related to Japan's Meiji Industrial Revolution. At present, the stone factory is utilized as a museum, Shoko Syusei-kan, and the enormous documents reciting the history of Shimadzu's 700 year rule are stored there.

ヤマザクラ　*Prunus serrulata* var. spontanea f. humilis Wilson.　**祇園之洲**　Gion–no–su　**1914-03-14**

官軍墓地。樹高 7.6m 幹周 0.6m　後ろにクロマツ

Loyalists cemetery. Height 25ft. Circumference 2ft. *Pinus thunbergii* Parl. behind.

2016-02-03

黒々としたマツの木立の前で、白い輪郭を際立たせる八分咲きのヤマザクラ。そして、累々と立ち並ぶ西南戦争官軍将校の墓石群とあどけない少年のまなざし。静謐な風景のなかに、未来への希望を写し取ろうとした撮影者の思いが伝わる1枚である。

1945（昭和20）年7月、米軍機の空襲で一帯は黒煙に包まれた。10年後に官軍の遺骨は地下の納骨堂に合祀され、墓地は公園として整備された。「明治11年8月」と年号が刻まれた写真の石碑だけが戦禍を奇跡的に生き延びて、ぽつねんと公園の片隅に立っている。

Yama-zakura (*Prunus serrulata*) 80 percent in bloom. The deep grove of Japanese Black Pine in the background made the white outline of cherry blossoms conspicuous. A group of gravestones of Imperial Army Officers of the Satuma Rebellion are lined up and a boy with an innocent look stood still. It's a piece that reflects the photographer's thought of hope for the future in a tranquil setting.

In July of 1945, the whole area was surrounded by black smoke after air raids by U.S. military planes. 10 years later the ashes of the Imperial Army Officers were enshrined in the underground mausoleum, and the graveyard was developed into a park. Only the monument in the photograph, 'August, 1878' engraved upon it, which miraculously survived the war damage now stands alone in the corner of the park.

カンヒザクラ　*Prunus campanulata* Maxim.　**磯街道**　Iso Shore road　**1914-03-03**

花が下垂している。樹高 7.6m　幹周 0.6m

Flowers pendulous. Height 25ft. Circumference 2ft.

写真提供：仙巌園　Photo provided by Sengan-en.

「この愛らしさを表現する言葉は思いつかない」とウィルソンが絶賛した磯街道沿いに咲いていたカンヒザクラ。濃いピンクの花が下向きに咲く様子はまるで釣り鐘のようだと表現した。桜島大噴火の降灰で何もかもが灰色に染まって見える風景のなかで、復興の兆しを感じさせてくれるものであったに違いない。

春いちばんを告げるカンヒザクラは、石垣島、台湾などに自生する亜熱帯性のサクラで、早咲きの園芸種の親となっていることが多い。

Kanhi-zakura (*Prunus campanulata*) which was blooming along the Iso shore road. Wilson praised this Cherry by stating “I cannot think of words that could really describe this loveliness.” He described how the dark-pink flower bloomed downward just like a hanging bell. This must have made people feel a sign of restoration in scenery where everything looked gray because of the ash fall from of the great Sakurajima eruption.

Kanhi-zakura that first tells the arrival of spring is a subtropical tree that grows wild in areas such as Ishigakijima (one of the Okinawa islands) and Taiwan. It has been crossbred with different varieties as a parent for early-blooming garden plants.

採集日 Collection Date = 1914-03-03

海洋国家薩摩と琉球

曲水の庭に隣接して江南竹林がある。竹林奥に建つ石碑「仙巌別館江南竹記」によると、1736（元文元）年、島津家21代当主吉貴が琉球国に中国江南地方原産のモウソウチクを所望し、２株を取り寄せ移植したところ、領内に見事に根付いたとされる。

モウソウチクの日本伝来については、高僧に由来する京都起源説や、近い所では蒲生の旧領主が伝えたという蒲生伝承説など諸説存在する。だが、18世紀後半に江戸で大流行した記録があることから、徳川家とパイプを太くしていた島津氏が将軍に献上し、やがて庶民に広まったという仙巌園発祥説は説得力がある。春いちばんに芽を吹くモウソウチクの筍は、古来日本にあったといわれるマダケに比べ柔らかく美味しい。季節柄、ウィルソンも新鮮な筍の味覚を楽しんだ可能性はある。日本に先立つ中国探検で目にした江南の竹が、島津公爵の庭に繁茂している。その背景をも咀嚼しながら、カメラを向けた思いを想像してみる。

1609（慶長14）年、財政危機にあった薩摩が琉球国に武力侵攻し、中国への進貢貿易の利権を握った。砂糖の生産や中国から伝わる情報の入手など、江戸時代を通して琉球口と呼ばれるこの間接貿易で得た利益と海外情報が、斉彬の集成館事業へと集約されていった。琉球を窓口にモウソウチクが全国に拡大したように、薩摩から近代日本の礎となる技術が波及していったのである。

「望嶽楼」も19代島津光久の代に琉球の国王から献上されたもので、藩主が琉球使者と面接する際に使用された。幕末には、オランダ海軍の二等軍医ヨハネス・ポンペとともに長崎から幕府の軍艦ヤーパン丸（後の咸臨丸）に乗ってきた勝海舟と斉彬が会談を行った歴史的な場所でもある。

また、磯街道の海岸側の路肩には石灯籠とともに「琉球人松」と呼ばれる見事な枝を張った樹齢100年級の大マツが戦後まで立っていた。琉球からの船が入港する時に目印にしたという。害虫にやられて切り倒されてしまったが、現在は石燈籠の左手に姫松と呼ばれるクロマツの一種が数本、右手に那覇市から本土復帰１周年の記念に寄贈されたリュウキュウマツ２本が植栽されて育っている。

仙巌園を訪れた翌日、３月17日にウィルソンは次の探索地である長崎に向かう列車に乗った。そして３年後に再び日本の地を踏んだ時、真っ先に沖縄に足を運んでいる。仙巌園周辺で見た琉球の香りに、南の島へと興味を誘われたのかもしれない。

琉球人松（大正期の絵葉書）
Ryukyuan Pine on the shore（Postcard in early 20th century）.

同アングルから撮った現在写真　2015-11-30
View from the same angle.

Maritime Satsuma and Ryukyu

In Sengan-en, there is a bamboo grove adjacent to the garden of the winding stream. According to a stone monument in the grove, the 21st head of the household, Yoshitaka, desired Chinese Moso Bamboo (*Phyllostachys edulis*) from the kingdom of Ryukyu (now Okinawa) and planted two in the garden in 1736. They rooted wonderfully and soon spread to the whole of Japan.

In regard to the introduction of Moso Bamboo, various theories of origin exist. One is the Kyoto origin theory that said it was brought by monks dispatched to China and another is that a former lord of Kamo brought it from Ryukyu. However, since there is record that it was all the rage in Edo late in the 18th century; the Sengen-en origin theory seems most likely because Shimadzu family maintained close ties with the Tokugawa shogunate. The shoot of Moso Bamboo sprouts early in spring and it is much softer and more delicious than the Japanese Timber Bamboo (*Phyllostachys bambusoides*) which is said to have existed from ancient time. Since Wilson was visiting at that season, it's possible that he had enjoyed the taste of fresh Moso Bamboo shoot here. The Bamboo that Wilson saw during his earlier China expedition also grew thick in the garden of Shimadzu. Given that situation, I try to imagine his thoughts when he pointed his camera to the grove.

In 1609 the Satsuma clan was in financial crisis and invaded Ryukyu by force of arms, grabbing the concession of trade with China. The profit from production of sugar and the acquisition of information from China by this indirect trade through Ryukyu were consolidated in Nariakira's industrial activities. As Moso Bamboo had spread across the country, the technology that became the foundation of modern Japan began to spread from Satsuma.

'Bogaku-ro,' the Chinese styled pavilion was a tribute from the king of Ryukyu to the Shimadzu Clan and had been used for meetings with envoys from Ryukyu. It is also a historic site where Kaishu Katsu, a vassal of the Tokugawa Shogunate, coming from Nagasaki on the government's warship, the 'Yapan-maru,' with Johannes Pompe, surgeon of the Dutch Navy, met with Nariakira in 1858.

Another Ryukyu connection was a huge old Pine with wonderful branches called 'the Ryukyuan Pine,' which stood next to a stone lantern on the shoulder of the coastal highway until after the war. It was used as a landmark when ships from Ryukyu entered the port. The tree was cut down because of damage caused by pests, replaced by a couple of Black Pines on the left side of the stone lantern and two other Ryukyu Pines (*Pinus luchuensis*) standing on the right side. They were donated by Naha city, Okinawa in commemoration of the first anniversary of Okinawa's returning to Japanese sovereignty from the reign of the U.S. in 1973.

The day after visiting Sengan-en, Wilson boarded the train to Nagasaki, the next point on his expedition. When he set foot on Japan three years later, he went to visit Okinawa first. I surmise that a fragrance of Ryukyu which he saw around Sengen-en piqued his interest in the southern island.

琉球国王献上の楼閣「望嶽楼」 2015-11-30
Bogaku-ro, gift of the king of Ryukyu.

藩主が琉球使者と面接する際に使用したといわれている東屋。床には、秦の阿房宮の床瓦を模造したと伝えられる床瓦が273枚敷きつめてある

Pavilion which was used when the Shimadzu lord met with envoys from Ryukyu. With 273 tiles covering the floor, it imitates the palace of the Qin dynasty in China.

ナギ　*Podocarpus nagi* Zoll. & Moritz.　**仙巌園**　Sengan-en　**1914-03-16**

島津公爵の庭にて。樹高 15m 幹周 1.2m
Garden of Prince Shimadzu. Height 50ft. Circumference 4ft.

2015-03-28

「曲水の宴」で使われる島津家の紋が入った杯
Sake cup with the Shimadzu's family crest, which was used for "Kyokusui-no-en."

針葉樹でありながら、広葉樹のように幅の広い葉を持つナギの姿はウィルソンの目には珍しかったようだ。葉が光り、まっすぐに伸びることから精霊が宿る木として寺社に植栽されることが多く、また凪(なぎ)に通じるところから、航海の平穏を祈る神木ともされてきた。この木の存在もまた、海洋国薩摩のひとつの表象であったのかもしれない。

現在も、清水の流れを見下ろす場所に立っている。その足元で毎年春になると「曲水の宴」と呼ばれる雅な歌会が催される。ナギの周辺にスギも植えられているが、台風の影響で傾いている。

The appearance of Nagi (*Podocarpus nagi*) having wide leaves like a broadleaf tree while being a conifer seemed to be rare to Wilson's eyes. Because of its shiny leaves and straight trunk, it is often planted in temples and shrines as the tree in which spirits dwell. Since Nagi can also mean 'calm sea,' it was considered a sacred tree to which seafarers pray for the safety of a voyage. The existence of this tree might be a representation of the maritime nature of Satsuma.

The tree still stands in the place where it can overlook a brook. Every year in the spring, the poetry party called 'Kyokusui-no-en' is held at the base of the tree. Cedar trees are planted around the Nagi, but one is leaning as a result of typhoon.

モウソウチク　*Phyllostachys mitis* Riv.　**仙巌園**　Sengan–en　**1914-03-16**

1736年、島津公爵の祖先により琉球からもたらされた竹林の一部。島津公爵の庭で今も繁茂している
Part of original grove brought from Liukiu in 1736, by ancestor of prinse Shimadzu in whose garden at Kagoshima it is now growing.

2015-03-28

モウソウチクの発祥の地としての謂れを記した「仙巌別館江南竹記」の碑
Monument that states the origin of the birthplace of the Moso Bamboo (Sengan Bekkan Konan-chiku-Ki).

1736（元文元）年、島津家21代当主吉貴が琉球経由で2株取り寄せたという石碑の記録が残る江南竹林と呼ばれるモウソウチクの林。今や全国に広がったモウソウチクの発祥の地とされる。撮影当時はスギが近くに植えられていたが、モウソウチクの侵入でスギが弱っているのが見てとれる。

1993（平成5）年の鹿児島8・6水害の際に、集中豪雨による崖崩れで流失。現在の竹林はその3年後に再移植されたものである。海からの風が奏でる笹音と繊細で軽快な光が演出する清涼感は、訪れる人の五感を楽しませてくれる。

The grove of Moso Bamboo (*Phyllostachys edulis*) which is called 'Konan Bamboo Grove.' According to the stone monument the Shimadzu's 21st household, Yoshitaka, ordered two stocks of them from China via Ryukyu. This place is considered to be the origin of Moso Bamboo in Japan, which has spread throughout the country. When Wilson took the photograph there were several Cedars in the grove. They looked weak because of the invasion of Bamboo roots.

This grove was washed away by a landslide caused by torrential rain on August 6, 1993. The current grove was formed three years later with Bamboo replanted from the disaster. The sound of bamboo leaves played by the sea breezes and the delicate flickering light please the senses of visitors.

column 殿さまの庭のヤクタネゴヨウ
Yakutane-goyo in the Garden of a Feudal Lord

仙巌園の正門から入って右側、御殿前庭で身を傾けながらも守護神のように庭園を見守っているマツの木は、樹齢350年といわれるヤクタネゴヨウの老木である。地球上で屋久島と種子島にのみ自生する希少種で、環境省のレッドデータブックで「絶滅危惧種」に指定されている。

このマツに最初に関心を寄せたのがウィルソンであったことは、意外と知られていない。

「鹿児島の島津公爵の庭で、種子島で出逢ったものと同じ五葉松を見た。私はタカネゴヨウではないかと思うが……」

確かに中国南部で目にしたタカネゴヨウと似ているが、球果や葉の様子が異なり変種ではないかという疑問を論文で展開している。

ウィルソンが最初にこのマツに出逢ったのは屋久島に向かう途中で、船がわずか数時間、種子島に寄港した時だった。瞬時にこの地域固有の希少なマツであることを見抜いたが、球果がついていなかったので新芽だけを乾燥させた。この時点で、他の植物学者は誰もまだその特異性に気づいていなかった。その後、タカネゴヨウの変種であることが確定され、1938（昭和13）年に分布地にちなんでヤクタネゴヨウと名付けられた。

かつては単に五葉松と呼ばれ、丸木舟の用材として盛んに利用されていた。過度の伐採や害虫の被害で個体数が減少し、現在では屋久島の世界自然遺産地域内の西部林道沿い斜面などに白骨化した樹幹を累々とさらすものもあり、地元有志が熱心な保護活動を進めている。

ウィルソンが訪れた時には仙巌園にも十数本あったのが、今はその1本だけが生き残っている。写真こそ撮る余裕がなかったが、この仙巌園の庭のマツはウィルソンの記憶に強く残った。後日、「東京で牧野富太郎からこの庭で採集した球果と枝をもらった」と、新しい玩具を与えられた子供のように無邪気な喜びを記録している。

When entering from the front gate of Sengen-en, on the right in the palace front yard, you'll see a 350-year-old Yakutane-goyo (*Pinus armandii* Franch. var *amamiana*), inclining it's body and watching the garden like a guardian deity. It is a rare species that grow wild only in Yakushima and Tanegashima, and is designated 'an endangered species.'

It is not surprising that it was Wilson who first showed special interest in this pine. He wrote in his book "I saw growing in a garden of Prince Shimadzu a white pine which looked to me like *P.armandii*. This suggested the idea that the Goyo-matsu of Tanegashima and Yakushima might belong here."

Wilson saw this Pine for the first time when a few weeks earlier his ship stopped at the port of Tanegashima. He noticed a Pine resembling one he saw in China, but immediately suspected it to be a variant indigenous to this area, observing differences in the cone and needles. At this point, no other botanists had noticed this specificity. In 1938, it was determined to be a variety of *P.armandii*; was named 'Yakutane-goyo,' thus confirming Wilson's theory.

It was once called merely Goyo-matsu (five-needled-pine) and was used in making dugout canoes. Its population is decreasing due to excessive felling and damage from insect pests. Currently, there are many exposing their skeletal trunks on the slopes along the western part of the World Natural Heritage Area of Yakushima. Local volunteers are actively making effort to protect remaining trees.

When Wilson visited Sengen-en, there were a dozen of them, but only one survives. He did not take a photograph, but it remained in his memory. Later on he recorded his joy like when child is given a new toy; "In Tokyo, T. Makino gave me cones and a branch, collected in the garden of Prince Shimadzu."

ヤクタネゴヨウ *Pinus armandii* Franch. var *amamiana* (Koidz.) Hatus.
仙巌園 Sengan-en **2015-11-30**

樹高 21m 幹囲 5.4m 推定樹齢 350 年

Height 70ft. Girth 18ft. Estimated age 350 years.

仙巌園内の古木の１本。長い年月のあいだに幹の内部が空洞となり、西に傾いている。現在は、枝の重みや台風によって倒れないように添え木とワイヤーロープで固定してある。定期的に、台風などの風圧を軽減するため剪定が行われる。

ちなみに、屋久島に自生する樹高 30m、幹周 6.8m の個体が、現在確認されているヤクタネゴヨウの最大木とされている。

One of the old trees in Sengen-en. The inside of the trunk became hollow and inclined to the west after long time. It is now fixed by supporting poles and wire rope to protect it from the heaviness of its branches and typhoons. On a regular basis, a pruning operation is conducted to reduce the wind pressure from typhoons.

The largest known Yakutane-goyo (*Pinus armandii* Franch. var *amamiana*) height 30m, girth of 6.8m, is growing wild in Yakushima.

E.H. ウィルソン　標本コレクション

ウィルソンのフィールド・ノートの記録から、1914（大正3）年2月から3月に鹿児島を訪れた際に、289点を数える植物を採集し乾燥標本を作成していたことがわかった。1917（大正6）年の2回目の日本訪問と合わせて、沖縄、小笠原から北海道、サハリンまで日本列島をくまなく訪れ、合計2354点の標本を採集している。写真については全画像データを入手することができたが、標本は散逸してしまっているであろうと考えていた。ところがハーバード大学標本館を訪れた際に、100年前のウィルソンの仕事を物語る実物標本の多くが素晴らしい保存状態で残されているのを確認した。

大学の標本館としては世界最大と言われるハーバード大学標本館（1842年設立）は、19世紀中期から現在までの約500万点におよぶ標本を所有している歴史ある施設である。データベース化はまだ完成していなかったため、学名ごとに分類されたキャビネットの中からウィルソンが日本で作成した標本を見つけ出すのは時間のかかる作業ではあったが、学芸員の手を借りて、全部で158点の標本を見つけ出すことができた。

標本の右下に貼られたラベルには、ウィルソンが自筆で植物名、日付、採集場所、標高、繁殖状況などを書き記している。添えられた番号が、フィールド・ノートの番号と完全に一致することから、彼の仕事の緻密さと植物学者としての真摯な姿勢を改めて実感した。また日付、採集場所が写真と見事にマッチングし、モノクロの写真に色彩イメージを与えてくれるものもいくつかあった。

まるで、たった今、採集したばかりのようなみずみずしい色合いを保っている葉や花や種の現物を間近に見た時、これらを保管しながら旅を続け、ボストンへ蒸気船で送る際にも厳重に梱包したであろうウィルソンの努力、そして、その努力を損なうことなく100年という年月の間、適切な環境で管理し続けてきたハーバード大学標本館の姿勢に感服せざるを得なかった。

標本も写真と同様に植物学的な意味合いだけでなく、歴史そのものを記録するドキュメンタリーとして貴重なものである。

発掘することができたウィルソンの日本での植物標本のうち、屋久島と鹿児島本土で採集されたもの18点をピックアップしてここに並べてみた。なお、本文で紹介したものは割愛した。

1917年と1918年の探検でウィルソンが集めた植物標本。ボストン、アーノルド樹木園フェンネル管理棟1階の標本アトリウムで撮影(1919年4月)

E.H. Wilson's collection of herbarium specimens made in 1917 and 1918. Photographed in the atrium,1st floor, herbarium wing, Hunnewell Administration Building at Arnold Arboretum in Boston, April 1919.

Wilson's Herbarium Specimens

I learned from the record of his field notebook that Wilson collected 289 plants, preserved as herbarium specimens, when he visited Kagoshima from February to March in 1914. He collected a total of 2354 specimens in Japan during his two visits in 1914 and 1917, from Okinawa to Sakhalin. I have been able to obtain image data for all of his photographs, but I worried that his specimens could be scattered or lost. However, when I visited the Harvard University Herbaria, I saw to my relief that the specimens from Wilson's own hands from 100 years ago have been preserved in splendid condition.

The Herbaria, founded in 1842 is the largest university owned herbarium in the world with over 5 million specimens dated from mid-19th century to now. Because computerized data bases are not yet completed, it was a time consuming task to find the specimens Wilson made in Japan from the maze of cabinets which are arranged by scientific name, but with the curator's help, I was able to find 158 of them.

The plant name, date, collection place, altitude and the growing conditions were written on a label affixed to the lower right of the specimen in Wilson's handwriting. The number written on the label perfectly matched with the number in his field notebook. I was impressed anew by the attentiveness to detail in his work and his thoroughness as a botanist. In addition, the date of the specimens matched the photograph date wonderfully, so some monochromatic photographs could be given a color image.

When I observed the specimen of a leaf, a flower or seed which kept the freshness and youthful hue as if they were just collected, I could not help admiring Wilson's efforts carrying them with great caution and packing them so carefully when shipping to Boston by steamship. Just as wonderful is the care given by Harvard University Herbaria to preserve Wilson's efforts perfectly in the appropriate environment over the past 100 years.

Like his photographs, Wilson's specimens are precious, not only for botanical reasons but also as valuable historical artifacts. On next page, I introduce another 18 specimens which Wilson collected in Kagoshima.

ハーバード大学構内にある標本館
The Herbaria in the campus of Harvard University

横浜からボストンに船便で送られたウィルソンの標本の荷札。1919（大正8）年1月
A shipping tag of Wilson's specimens from Yokohama to Boston in Jan, 1919.

写真提供：ロイ・ブリッグス
Photo provided by Roy W. Briggs.

クロキ 屋久島 Yakushima　1914-02-18
Symplocos lucida S. & Z.

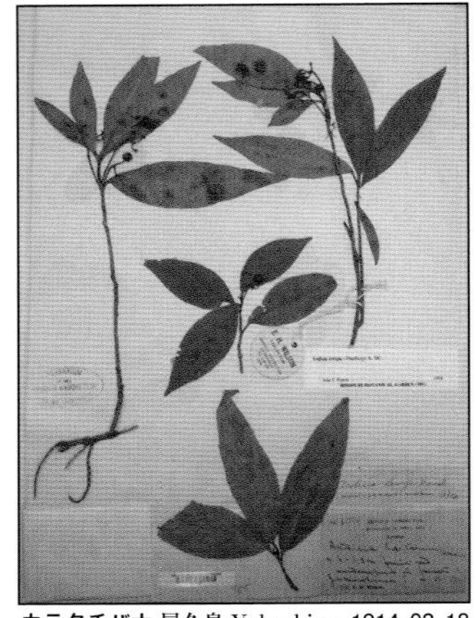

カラタチバナ 屋久島 Yakushima 1914-02-18
Ardisia crispa（Thunb.）A. DC.

ヒカゲツツジ 屋久島 Yakushima 1914-02-19
Rhododendron keiskei Miq.

ヤシャブシ 屋久島 Yakushima　1914-02-19
Alnus firma S. & Z.

ナナカマド 屋久島 Yakushima　1914-02-24
Sorbus commixta Hedl.

オガタマノキ 屋久島 Yakushima 1914-02-24
Michelia compressa Maxim.

ノリウツギ 屋久島 Yakushima 1914-02-24
Hydrangea paniculata Siebold

ウバメガシ 屋久島 Yakushima 1914-02-25
Quercus phillyreoides A.Gray

サキシマフヨウ 屋久島 Yakushima 1914-02-25
Hibiscus mutabilis L.

＊学名はウィルソンの表記のまま　Based on the scientific name Wilson used.

スイカズラ　屋久島 Yakushima 1914-02-25
Lonicera japonica Thunb.

スダジイ　屋久島 Yakushima　1914-02-26
Castanopsis cuspidata Schottky

シロダモ　鹿児島市内 Kagoshima City
Litsea glauca S. & Z.　1914-03-02　※

コマツナギ 鹿児島市内 Kagoshima City *Indigofera pseudotinctoria* Matsum.　1914-03-03

ヤマヤナギ　鹿児島市内 Kagoshima City
Salix sieboldiana Blume　1914-03-03

ヤマグルマ　霧島 Kirishima　1914-03-11
Trochodendron aralioides S. & Z.

クロキ　川内 Sendai　1914-03-14
Symplocos lucida S. & Z.

アケビ　川内 Sendai　1914-03-14
Akebia quinata Dence.

イロハモミジ　磯 Iso　1914-03-16
Acer palmatum Thunb.

※ 現在の学名（Present Scientific name）：*Neolitsea sericea*

プラント・ハンターの必需品

日本での旅でウィルソンはしばしば、列車に持ち込む荷物に別途料金が課され、それが度重なると無視できないほど大きくなることに愚痴を述べている。写真撮影と同時に、旅を続けながら採集植物の乾燥標本づくりも行っていたウィルソンは、撮影機材の他にどのような備品を常時、持ち歩いていたのだろうか。

残されたさまざまな資料から考察してみる。時計、コンパス、晴雨計、万歩計、携帯高度計といった基本的な機材の他に、生きたままあるいは標本にする植物を採集し保存するための道具があった。

まずは、採集した植物を入れる金属の箱（胴乱）や植物採集道具など。次に、最も大切なのは野外で植物を挟んで持ち歩くための携帯用簡易プレス（野冊）で、ウィルソンは中国で作らせた特別仕様のものを持っていた。

「9個のプレス機を使い、何千もの乾燥用紙を駆使して作業した。毎晩、給水紙を取り替えるのは大変な作業であるのは確かだ」

標本の品質劣化を最小限にするためには、可能な限りすばやく植物から水分を取り除くことが求められる。本格的な標本作りは主に、探検が終わり宿舎に戻った夜か、天候が悪くて外出できない日に行った。標本台紙、植物名を記し分類するためのラベル、日付と番号が入ったゴム製スタンプ、標本やラベルを台紙に貼りつけるための接着剤、インク、段ボール、紙、紐などが必要だった。集めた植物に標本づくりを追いつかせるのは、実に骨の折れる作業だったという。

乾燥させ、分類ラベルを貼ると、保存のため油紙でしっかりと梱包し、横浜まで持ち歩いた後に船便でボストンに送った。さらには、生きた植物を搬送させるための密閉型の「ウォードの箱」もいくつか携帯していたに違いない。1842年ロンドン在住の医者ナサニエル・ウォードなる人物が、密閉したガラス瓶のなかでシダの胞子が発芽し成長することから考案したもので、この画期的な植物輸送箱の発明により、プラント・ハンターたちは長い船旅でも生きた植物を安全に運ぶことができるようになった。

そして、もちろんウィスキーやブランデーも。無聊を慰めるだけでなく、焼酎や日本酒と交換しコミュニケーションをはかるための必携品でもあったことだろう。

これら数々のアイテムが、ウィルソンの旅の装備品だった。プラント・ハンターの仕事は英雄的で華やかに見える一方、日常の地道な努力にも支えられていたのだ。

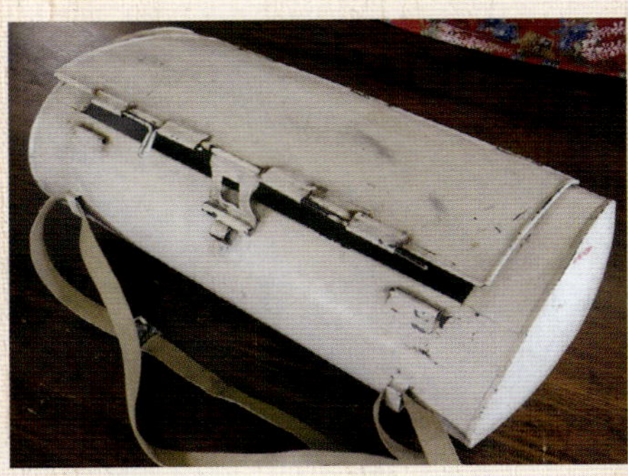

胴乱 Vasculum
採集植物を入れて肩から下げる植物採集用のブリキ製の携帯箱
Carrying box made of tin for plant collecting. Field samples were put in this box and hung over the shoulders of the plant hunter for transportation.

野冊 Fierld press
採集植物をその場で挟んで持ち歩くのに使用。吸水紙を挟む格子の板2枚とそれを縛る紐でできている
Used in the field to hold collected plants. It consisted of two frames of lattice boards, which held plants between absorbent paper, and straps to tie the boards tightly together.

The Necessities of a Plant Hunter

Traveling around Japan, Wilson often complained about the surcharge for baggage carried on the trains, which was difficult to ignore as it would add up to a great deal as he traveled. In addition to taking photographs, Wilson was busy collecting plants and preserving specimens during his expeditions. Beside his photographic equipment I wondered what other kind of supplies he had to carry.

Researching various documents I learned that beside the basics such as watch, compass, barometer, and altimeter, Wilson carried a huge amount of equipment to collect and preserve plants.

First of all was a Vasculum, a metal box he took in the field to carry back collected plants and tools to cut the plants. Next and most important was the Field Press to press the plants in the field and begin drying the specimens. Wilson had these made especially in China.

"I worked with 9 presses and a thousand 'Driers.' And I assure you changing specimens every night meant work." Wilson wrote.

In order to minimize the degradation of the specimen's quality, it was required to remove moisture from the plant as fast and as much as possible. He prepared specimen panels mainly at night or on days of bad weather. For this he needed specimen panels, identification labels, adjustable date and number stamps, adhesive for mounting specimens and labels, cardboard, string, ink. He had to work hard preparing the specimen panels as he was collecting new plants every day.

The completed specimens were carefully wrapped in oiled paper for shipment. In addition, he must have carried several airtight glass boxes called the 'Wardian Case.' These were invented in 1842 by Nathaniel Ward, a London physician, as a means to carry live plants safely even on long sea voyages.

Finally, of course, bottles of whiskey and brandy; they were indispensable not only to relieve boredom but also to trade for local Shochu and Sake, and as an aid to communication.

These were items Wilson required wherever he went. The work of the plant hunter may seem glorious, even heroic, but it required continuous hard work every day.

ウォードの箱　Wardian case

イギリス人ウォードが19世紀中に発明した画期的な植物運搬用の密閉箱。中の植物が光合成を行うことで発生した熱を逃がすため水を蒸散させ、それが水滴となって根が吸収するという水の循環を利用している
A sealed glass case for transporting live plants, utilizing water generated by photosynthesis of plants inside.

ウィルソン氏からの手紙

イギリスから一通の手紙が届いた。Eメールという電子伝達ツールを使うことが多くなった昨今、エリザベス女王の顔が印刷された切手が貼られ、ロンドンの消印が押された正真正銘の国際郵便が届くことは珍しく、ふんわりと温かいものに触れたような嬉しい驚きがあった。

中を開いて、さらに驚愕した。ミミズの這ったような判読困難な英文の羅列。それは、私にとっては、実に見慣れた文字であったからだ。

「この文字を見て、すでにもう誰か気づかれたかもしれません」という一文で手紙は始まっていた。そう、それは日夜、私が解読に汗水をたらしていたウィルソンの筆跡そのものであった。

この謎めいた手紙の差出人は、ロイ・ブリッグスという人物だった。アーネスト・ウィルソンの弟であるディック・ウィルソンの孫にあたる人で、2013年にウィルソンの伝記を出版した作家でもある。それにしても悪筆はかくまで遺伝するものだろうか。ようやく読みこなした手紙には、屋久島での人間ウィルソンを描いた私の本への感謝の言葉が述べられてあった。それは、ウィルソン本人から送られた手紙そのもののように私には思え、胸を熱くして手紙を抱きしめた。

2014年6月、ウィルソンの生まれ故郷チッピング・カムデンを訪れる機会に恵まれた。産業革命以前の街並みが残る美しい通りの一画にウィルソンが世界から集めた植物が植栽されている「メモリアル・ガーデン」があり、ハンカチノキが白い苞葉をひらひらと振って私を出迎えてくれた。その庭の管理事務所に置いてきた拙著が巡り巡って、ロイの目に触れたのであった。

ウィルソンが100年前に撮影した写真は、人と人をつなぐ不思議な力があるらしい。それから、イギリスはもちろんアメリカ、中国、台湾と各国のウィルソニアンたちとの情報交換の輪が水の波紋のように広がっていった。イギリスでウィルソンのドキュメンタリー番組を制作中のスーザン・ユーが、定期的に届けてくれる情報も貴重なものだった。

気がつけば、鹿児島においても画像を手に走り回る私とカメラマンの周りには、同じように写真の魅力に引きこまれていく仲間がひとり、またひとりと増えていった。新聞連載の最終章を手掛ける頃には、2台の車に分乗して撮影場所に向かうほどの人数となっていた。こうしていつしか「チーム・ウィルソン」が生まれ、多くの人々の協力に支えられて鹿児島県内のウィルソンの足跡をほぼ完璧に辿ることができた。

綿密な取材と研究に基づくロイの本は素晴らしい労作だが、ウィルソンの日本の旅については詳しくは書かれていない。地名も風景もずいぶん異なってしまっているので、在住の人間でない限り調査は困難だろうと想像する。だからこそ、微力ながら私がその空白を少しでも埋めることができれば、2人のウィルソンに喜んでもらえるのではないだろうかと勝手に思い込んでいる。

最後になったが、植物の特定および植生の変化などについて、ウィルソンの目となってアドバイスを下さった鹿児島県立博物館学芸員の寺田仁志氏に心から感謝したい。

今後も「チーム・ウィルソン」の輪が地球のあらゆる場所に広がっていくことを願って。

A Letter from Mr. Wilson

Last year, I received a letter from England. Since the electronic transmission tool called E-mail became a common method of exchanging messages, it was very rare to receive a real international letter, postmarked London, with a postage stamp of Queen Elizabeth. It was a delightful and heartwarming surprise.

I was even more surprised when I opened the letter. There were obscure English sentences crawling like earthworms, in a penmanship very familiar to me.

"Perhaps you've already noticed who I am just by looking at my handwriting," the letter began. Indeed, it was the handwriting of Wilson himself that I had gone to great length to decipher day and night.

The sender of this mysterious letter went by the name of Roy W. Briggs. He was the grandchild of Dick Wilson who was the younger brother of Ernest Wilson and was also a writer who published Wilson's biography, " 'Chinese' Wilson" in 2013. I cannot help but wonder if difficult handwriting is inherited. In the letter that I finally comprehended, his words of gratitude for my last book about Wilson in Yakushima were even more gratifying to me. It felt so much as if the letter was sent by Wilson himself that my heart was filled with joy and I pressed the letter to my heart.

In June of 2014, I was blessed with the opportunity to visit Chipping Campden where Wilson was born. There is 'The Memorial Garden' in this beautiful pre-industrial English town, where plants Wilson collected from all over the world are cultivated. I was welcomed by the Handkerchief Tree (*Davidia involucrata*) waving its white bract. I left my book at the management office of the garden, which was passed on from one to another, and eventually caught Roy's attention.

The photographs Wilson took 100 years ago seem to have a mysterious power to connect people. Recently the exchange of information among 'Wilsonians' between countries like the U.S., China, Taiwan, as well as the U.K, has expanded like ripples on water. Susan Yu, who is producing a documentary film on Wilson in the U.K., has dispensed valuable information regularly as well.

Before I knew it, even in places like Kagoshima, people who were also spellbound by Wilson's photographs started to gather around us one by one on a journey that had begun with a cameraman and me running around with picture images in our hands. By the time I was writing up the last column of the newspaper series, the size of the group had become so large that we divided the people into two cars to head for shooting locations of the photographs. In this way, 'Team Wilson' was eventually formed, and with the support of many, I was able to trace Wilson's footprints nearly to perfection.

The book for which Roy had done extensive research is a splendid laborious work, but little was written about Wilson's trip to Japan. Because the name of places and the scenery have changed so much, I imagine it would have been difficult to investigate that trip unless done by a local resident. Therefore, I felt I would be able please the two Wilsons if I could help fill in that missing gap in Wilson's story.

I hope the circle of 'Team Wilson' will continue to spread all around the world.

主な参考文献　References

日本文資料　Japanese sources

白幡洋三郎『プラントハンター』講談社学術文庫、2005年
F.キングドン - ウォード、塚谷裕一訳『植物巡礼―プラント・ハンターの回想』岩波文庫 、1999年
アリス・M・コーツ、遠山茂樹訳『プラントハンター東洋を駆ける―日本と中国に植物を求めて―』八坂書房、2007年
ロバート・フォーチュン、三宅馨訳『幕末日本探訪記 江戸と北京』講談社学術文庫、1997年
H.B.シュワルツ、島津久大・長岡祥三訳『薩摩国滞在記 宣教師の見た明治の日本』新人物往来社、1984年
T.ホイットル、白幡洋三郎・白幡節子訳『プラント・ハンター物語 植物を世界に求めて』植物と文化双書、八坂書房、1983年
田代善太郎『鹿児島県屋久島の天然記念物調査報告（復刻版）』（有）生命の島、1995年
田代晃二編著『田代善太郎日記 大正編』創元社、1972年
山本秀雄編著『屋久島歴史小年表』（有）生命の島、2007年
太田五雄編著・発行『屋久島山岳体系（五高山岳部復刻版）』、2013年
金谷整一『屋久島自然系 その1 絶滅の危機にあるヤクタネゴヨウ』生命の島第52号、（有）生命の島 、2000年
上屋久町郷土誌編集委員会『上屋久町郷土誌 全一巻』、1984年
屋久町郷土誌編さん委員会『屋久町郷土誌 第三巻 村落誌下』、2003年
杉崎元『ウイルソン株』暖友林7月号、熊本営林局林友、1956年
金平亮三『ウイルソン君を懐ふ』台湾博物学会報第21巻113号、台湾博物学会、1931年
初島住彦監修『屋久島高地の植物』南方新社、2001年
初島住彦監修『世界自然遺産の島 屋久島の植物』八重岳書房、1995年
（公益財団法人）屋久島環境文化財団『屋久島の植物ガイド』、2005年
（公益財団法人）屋久島環境文化財団『屋久島環境文化村ガイド 図説屋久島』、1996年
高知新聞社編『MAKINO ―牧野富太郎生誕150年記念出版―』北隆館、2014年
上村登『牧野富太郎伝』六月社、1955年
牧野富太郎『牧野富太郎自叙伝』長嶋書房、1956年
牧野富太郎・武井近三郎『牧野富太郎博士からの手紙』高知新聞社、1992年
牧野富太郎『牧野富太郎選集』東京美術、1981年
大田猛彦『森林飽和 国土の変貌を考える』NHKブックス［1193］、NHK出版、2012年
鹿児島県立博物館編『大正三年 桜島大噴火写真集』鹿児島県教育委員会、1988年
鹿児島市役所『郷土教育資料 鹿児島地誌』、1935年
鹿児島県高等学校歴史部会編『鹿児島県の歴史散歩』山川出版社、1992年
南日本新聞社編『郷土紙にみる 世相百年』南日本新聞開発センター、1981年
南日本新聞社編『鹿児島百年（下）大正昭和編』、1968年
芳 即正編『鹿児島県民の百年―明治から昭和へ』鹿児島歴史シリーズ6、著作社、1987年
原口泉他『鹿児島県の歴史』山川出版社、1999年
唐鎌祐祥『昔の鹿児島―かごしま新聞こぼれ話』南日本新聞開発センター、2008年
五代夏夫『桜島の顔』高城書房出版、1989年
下薗三州児編『新鹿児島』新鹿児島編纂所、1915年
鹿児島市編『鹿児島のおいたち』丸山学芸図書、1984年
鹿児島商工会議所70年史刊行委員会編『鹿児島商工会議所70年史』鹿児島商工会議所、1965年
宮内勝典『火の降る日』河出書房新社、1983年
久木田末夫『鹿児島の鉄道・百年』かごしま文庫64、春苑堂出版、2000年
下堂園純治『かごしま歴史散歩』南洲出版、1977年
佐藤 剛『ふるさと歴史散歩（鹿児島）』地方・小出版流通センター、2000年
川越政則『南日本風土記』至文社、1962年
岩中祥史『鹿児島学』草思社、2012年
鹿児島純心女子大学国際文化研究センター編『新薩摩学1「世界の中のさつま」』南方新社、2002年
難波経健『大正三年櫻島大爆震記』南日本新聞開発センター、2014年
大野照好『鹿児島の植物』かごしま文庫3、春苑堂出版、1992年
豊増哲雄『古地図に見るかごしまの町』かごしま文庫30、春苑堂出版、1996年
唐鎌祐祥『天文館の歴史』かごしま文庫5、春苑堂出版、1992年
鹿児島の自然を記録する会編『川の生きもの図鑑』南方新社、2002年
寺田仁志『日々を彩る 一木一草』南方新社、2004年
鹿児島県『鹿児島県史蹟名勝天然記念物』、1937年
林 吉彦『鹿児島之史蹟』、1930年
竹村民郎『大正文化帝国のユートピア』三元社、2010年
五代秀尧・橋口兼柄編『三国名勝図檜 上・中・下巻』山本盛秀、1905年
鹿児島市教育委員会編『鹿児島文化財の手引き』市民運動推進協議会、1991年
鹿児島県郷土資料研究会編著『史跡と人物でつづる鹿児島県の歴史』光文書院、1979年
原口泉監修『図説 薩摩の群像 歴史群像シリーズ』学習研究社、2008年
今吉弘・徳永和喜編著『鹿児島県謎解き散歩』新人物文庫、新人物往来社、2012年
毎日新聞鹿児島支局編『城山物語』、1969年
第七高等学校造士館編『第七高等学校造士館一覧 大正4年9月―大正5年8月』、1916年
五代夏夫『薩摩秘話』南方新社、2002年
島津斉彬・牧野伸顕序『島津斉彬言行録』岩波文庫、1944年
國光社編『照国公感旧録』静思社、1899年
西元肇『照国神社境内の史跡・探勝園内の史跡』、2001年
鹿児島市中学校理科部植物研究グループ編『鹿児島市城山公園植物調査報告別冊』鹿児島市役所自然愛護課、1966年

中村明蔵『薩摩民衆支配の構造』南方新社、2000 年
姶良町歴史民俗資料館『特別展図録「越前（重富）島津家の歴史」、2004 年
鹿児島県蒲生町『蒲生郷土誌』、1969 年
日高一雄『蒲生の歴史と産業』蒲生町、1981 年
国指定特別天然記念物「蒲生のクス」保護増殖事業報告書編集委員会編『国指定特別天然記念物「蒲生のクス」保護増殖事業報告書』蒲生町教育委員会、2000 年
串間俊文『鹿児島の園芸植物』かごしま文庫 26、春苑堂出版、1995 年
松下重資編『鹿児島県郷土誌系統史』島津編輯所、1930 年
『目で見る国分・姶良の 100 年』郷土出版社、2004 年
姶良市歴史民俗資料館編『蒲生八幡神社の歴史』、2011 年
窪田仲市郎『霧島神宮』かごしま文庫 28、春苑堂出版、1995 年
学習研究社『霧島神宮・鹿児島神宮 神話のふるさと』、2003 年
霧島町郷土誌編集委員会『霧島町郷土誌』霧島町、1992 年
窪田仲市郎『神話と霧島』講談社出版サービスセンター、1972 年
霧島神宮奉賛会編『霧島神宮沿革史』霧島神宮奉賛会
霧島館編『霧島硫黄谷温泉』
紀元 2600 年鹿児島県奉祝会編『神代並神武天皇聖蹟顕彰資料第 3 輯官幣大社霧島神宮』、1939 年
高尾あゆみ・伊東龍一『近世後期における霧島神宮の境内の景観』日本建築学会大会学術講演梗概集、2008 年
赤司辨蔵『久留米躑躅誌 第五版 初版明治 38 年』久留米赤司廣樂園、1934 年
梅野満雄『筑紫商工先達小傳』久留米商業学校調査部、1935 年
新田神社社務所編『国幣中社新田神社略誌』、1938 年
新田神社社権禰宜砥錦茂全『薩摩国総鎮守新田神社』、2010 年
九州歴史資料館編『薩摩川内新田神社』、1978 年
上東郷村編『菅原神社案内 藤川天神』
宮脇 昭『鎮守の森』新潮文庫、2007 年
川内郷土史編さん委員会『川内市史』鹿児島県川内市、1980 年
東郷町郷土史編集委員会『東郷町郷土史』鹿児島県薩摩郡東郷町、1969 年
田村省三『尚古集成館 島津氏 800 年の収蔵』尚古集成館、2006 年
島津興業『磯の植物散歩』、1996 年
池田俊彦『島津斉彬公伝』岩波文庫、1994 年
鮫島志芽太『島津斉彬の全容』ぺりかん社、1988 年
加藤 蕙『島津斉彬』PHP 文庫、PHP 研究所、1998 年
上原兼善『鎖国と密貿易—薩摩藩の琉球密貿易』八重岳書房、1981 年
上里隆史『海の王国・琉球「海域アジア」屈指の交易国家の実像』歴史新書、洋泉社、2012 年
高良倉吉『琉球の時代—大いなる歴史像を求めて』ちくま学芸文庫、筑摩書房、2012 年
豊見山和行・高良倉吉編『琉球・沖縄と海上の道』街道の日本史 56、吉川弘文館、2005 年
徳永和喜『海洋国家薩摩』南方新社、2011 年
警察資料保存会『祇園之洲官修墓地写真帖』
鹿児島新聞 1914.3.01 ～ 3.16
古河古松軒『西遊雑記』日本庶民生活史料集成より抜刷・複製、1980 年

英文資料 English sources

E.H.Wilson. 1913. *A Naturalist in Western China with Vasculum, Camera and Gun.* Cambridge Library Collection, Cambridge University Press. New York.
E.H.Wilson. 1916. *The Conifers and Taxads of Japan.* Cambridge Printed at The University Press, Boston, Mass.
E.H.Wilson. 1916. *The Cherries of Japan*. Cambridge Printed at The University Press, Boston, Mass.
E.H.Wilson. 1917. *Aristocrats of the Garden*. Doubleday, Page & Company, New York.
E.H.Wilson. 1920. *The Romance of our trees*. Doubleday, Page & Company, New York.
E.H.Wilson & Alfred Rehder. 1921. *A Monograph of Azaleas.* The University Press, Cambridge, Boston, Mass.
E.H.Wilson. 1927. *Plant Hunting Volume I&II*. University Press of the Pacific Honolulu, Hawaii.
E.H.Wilson. 1930. *Aristocrats of the Trees.* Dover Publications, Inc. New York.
Roy W. Briggs. 1993.*"Chinese" Wilson; A Life of Ernest H. Wilson 1876-1930*. The Royal Botanic Gardens, Kew, U.K.
James Herbert Veitch. 1906. *Hortus Veitchii, A History of the Rise and Progress of the Nurseries of Messrs. James Veitch and Sons*. James Veitch & Sons Limited, Chelsea, U.K.
Edward I. Farrington. 1931. *Ernest H. Wilson, Plant Hunter*. The Stratford Company, Boston, Mass.
Gwen Bell. *E.H."Chinese" Wilson, Plant Hunter*. Seattle Washington Journal American Rhododendron Society.
Alice M. Coats.1969. *The Quest for Plants: a History of the Horticultural Explorers*. Studio Vista Limited, London.
Michael Tyler-Whittle. 1970. *The Plant Hunters*. London.
Robert Fortune. 2004. *Yedo and Peking: A Narrative of a Journey to the Capitals of Japan and China*. Ganesha Pub Ltd.
Carolyn Fry. 2009. *The Plant Hunters: The Adventures of the World's Great Botanical Explorers*. Carlton Publishing Group, London.
Toby Musgrave, Chris Gardner, Will Musgrave. 1998. *The Plant Hunters*: Two Hundred Years of Adventure and Discovery around the World. Seven Dials, Cassell & C. London.
E.H.Wilson. *Correspondence 1899-1930 from Japan, Feb. 1914-Jan. 1915*. Archives of the Arnold Arboretum of Harvard University, Boston.
E.H.Wilson. *Field Notes on collected plants and seed, Feb.1914-Jan.1915*. Archives of the Arnold Arboretum of Harvard University, Boston.
Peter J. Chwany. *E.H.Wilson, Photographer*. Archives of the Arnold Arboretum of Harvard University, Boston.
Alfred Rehder. *Ernest Henry Wilson*. Archives of the Arnold Arboretum of Harvard University, Boston.
Richard A. Howard. *E.H.Wilson as a Botanist*. Archives of the Arnold Arboretum of Harvard University, Boston.

謝辞　Acknowledgements

感謝を込めて（敬称略）Appreciation

ウィリアム・フリードマン（ハーバード大学アーノルド樹木園園長）
リサ・ピアソン（ハーバード大学アーノルド樹木園図書館館長）
マイケル・ドスマン（ハーバード大学アーノルド樹木園学芸員）
アンソニー・ブラック（ハーバード大学標本館学芸調査協力員）
デイビッド・ブフォード（ハーバード大学標本館学芸主事）
ミッシェラ・シュムール（ハーバード大学標本館学芸調査協力員）
キャロル・ジャクソン（イギリス、チッピング・カムデン歴史協会）
ロイ・ブリッグス（イギリス在住作家）
スーザン・ユー（イギリス在住フリーランス映像作家）
キャサリン・フォックス（アメリカ在住リサーチ専門家）
小野寺浩（公益財団法人屋久島環境文化財団理事長）
田村省三（尚古集成館館長）
三嶽豊・三嶽公子（NPO 法人かごしま文化研究所「文学サロン月の舟」）
ウィリアム・ブラワー（カバーデザイン＆英訳アドバイス ）
浅田浩子（英訳アシスト）

William Friedman, Director of the Arnold Arboretum of Harvard University
Lisa E. Pearson, Head of the Library and Archive of the Arnold Arboretum of Harvard University
Michael S. Dosmann, Curator of Living Collections of the Arnold Arboretum of Harvard University
Anthony R. Brach, Curatorial Assistant and Research Associate, Harvard University Herbaria
David E. Boufford, Senior Research Scientist, Harvard University Herbaria
Michaela Schmull, Research and Curatorial Associate, Harvard University Herbaria
Carol Jackson, Chipping Campden History Society
Roy W. Briggs, Author in U.K.
Susan Yu, Documentary Film Maker in U.K.
Kathleen Fox, Research Specialist in Boston U.S.A.
Hiroshi Onodera, Chairman of Yakushima Environmental Culture Foundation
Shozo Tamura, Executive Director and General Manager of Shoko Shusei-kan Museum
Yutaka Mitake &Kimiko Mitake, NPO Kagoshima Cultural Study Academy “Literature Salon Moon Boat”
William Brouwer, Cover Design & English Translation Adviser
Hiroko Asada, English Translation Assist

そして、ウィルソンについての情報、記憶を提供して下さったすべての方々へ
And all people who provided information and memories about Wilson

協力　Cooperation

鹿児島県立博物館	Kagoshima Prefectural Museum
公益財団法人屋久島環境文化財団	Yakushima Environmental Culture Foundation
公益財団法人カメイ社会教育振興財団	KAMEI Foundation for the Promotion of Social Education
九州ろうきん	Kyushu Rokin, Labor Bank
南日本新聞社	Minami Nippon Newspaper
KTS 鹿児島テレビ	KTS Kagoshima TV Station
鹿児島市	Kagoshima City
霧島市	Kirishima City
薩摩川内市	Satsumasendai City
始良市	Aira City
屋久島町	Yakushima Town
鹿児島県歴史資料センター黎明館	Kagoshima Historical Research Center, Reimei-kan
仙巌園	Sengan-en
尚古集成館	Shoko Shusei-kan
西本願寺鹿児島別院	Nishi-Honganji Temple, Kagoshima Branch
照国神社	Terukuni Shrine
霧島神宮	Kirishima Shrine
狭野神社	Sano Shrine
蒲生八幡神社	Kamo Hachiman Shrine
菅原神社（藤川天神）	Sugawara Shrine（Fujikawa Tenjin）
新田神社	Nitta Shrine
あかまつ荘（霧島）	Akamatsu-so（Kirishima）

チーム・ウィルソン　Team Wilson

植物・植生監修　寺田仁志（てらだ・じんし）

Total Plant Adviser　Jinshi Terada

屋久島生まれ。鹿児島県立博物館学芸員。文化庁文化財部非常勤調査員、環境省自然環境保全基礎調査植生調査部会検討委員など。

Born in Yakushima. Senior Research Scientist of Kagoshima Prefectural Museum. Cultural Properties Specialist of the Agency for Cultural Affairs, Committee Member of Natural Environment Conservation of Ministry of the Environment.

水流芳則（つる・よしのり）

Yoshinori Tsuru

元鹿児島県立博物館館長
Former Director of Kagoshima Prefectural Museum.

牧　市助（まき・いちすけ）

Ichisuke Maki

元屋久島教育長／ウィルソンの山岳ガイドの子息　Former Superintendent of Education in Yakushima. The son of Wilson's mountain guide, in 1914.

川越保光（かわごえ・やすみつ）

Yasumitsu Kawagoe

［鹿児島撮影］
川越亮事務所カメラマン

Photographer of Ryo Kawagoe Photo Office.
Photographed current scenes in mainland Kagoshima.

日下田紀三（ひげた・のりぞう）

Nirizo Higeta

［屋久島撮影］
屋久島在住フリーカメラマン／元屋久島町立屋久杉自然館館長

Freelance Photographer. Former Director of Yakushima Yaku-Sugi Museum. Photographed current scenes in Yakushima.

ウィリアム・ブラワー

William Brouwer

［標本撮影］
建築デザイナー・木工作家

Architectural Designer & Woodworker.
Photographed Wilson's specimens in Boston.

西川高司（にしかわ・たかし）

Takashi Nishikawa

［屋久島ロケーション協力］
屋久島森林保護員

Yakushima Forest Ranger. Located Wilson's sites in Yakushima.

斎藤ひかる（さいとう・ひかる）

Hikaru Saito

［写真展示］
鹿児島県立博物館学芸指導員

Curator of Kagoshima Prefectural Museum.
Display in Exhibition.

坂元幸一（さかもと・こういち）

Koichi Sakamoto

［霧島ロケーション協力］
霧島ジオガイド

Kirishima Geo Park Guide. Located Wilson's sites in Kirishima.

大迫秀世（おおさこ・ひでよ）

Hideyo Osako

［鹿児島ロケーション協力］
フリー編集者

Freelance Editor. Located Wilson's sites and transportation guide in Kagoshima.

徳永文子（とくなが・ぶんこ）

Bunko Tokunaga

［総合ヘルプ］
ラジオ・パーソナリティ

Radio Personality. Total assist and kept team Wilson fed in Kagoshima.

おわりに

ウィルソンは、イギリスが第二次産業革命を迎えた頃に生まれた。軽工業中心から、化学、電気、石油および鉄鋼の分野で技術革新が進み、重化学工業へと産業構造が移行しつつあった時代だ。バーミンガムで働き始めた時、そこはスモッグで太陽が見えなくなったブラック・カントリーと呼ばれる工業地帯と化していた。キュー・ガーデンで職を得た頃には、ロンドンの町も石炭燃料の利用により深刻な大気汚染に見舞われていた。同じ頃、ロンドンに留学していた夏目漱石は、「痰を吐きて見よ真黒なる塊りの出るに驚くべし」と滞在記に書いている。当時のイギリスには今の北京やニューデリーと同じような状況が生まれようとしていたのだ。

9年余かけて中国奥地で辺境の地で生きる人々に接した後に日本に足を延ばして、ウィルソンは「あまりに近代化の道を急ぎ過ぎている」との感想を残している。欧米列強に追いつくことを目標に、日清日露の戦争勝利で勢いづいた日本はひたすら工業化への道を邁進していた。同時に、足尾鉱毒事件や富山県のイタイイタイ病が深刻な公害問題として取り上げられ始めていた。

日本の旅でウィルソンのプラント・ハンターとしての仕事は終わりを告げた。1920年からはオーストラリア、タスマニア、ニュージーランド、シンガポール、マラヤ連邦、インド、セイロン、ケニア、南アフリカなどの植物園を巡る18カ月の旅に出た。そこで、彼が目にしたのは壊滅的な森林破壊の姿だった。

「自然のバランスが壊されようとしている。植物は我々動物にとっては有害な二酸化炭素で生きている。もし植物の世界が破壊されたら、私たちは生きていけないのだ。偉大な自然財産である木材の供給を近視眼的な欲望で喪失させてはならない。次の世代のために、今の世代が生きていることを思い出すべきである」

ウィルソンは環境破壊が人間に及ぼす影響についての懸念をメディアに向かって盛んに発言するようになった。地球温暖化という言葉こそ生まれていなかったが、二酸化炭素の充満による地球の危機を100年前にすでに予告していたのである。

現在の私たちはウィルソンの時代より、少しは賢くなったと胸を張っていえるだろうか？ 異常気象、生態系の破壊などウィルソンが警鐘を鳴らした問題が、ますます顕在化してきているのが現状である。

100年という節目にウィルソンが私たちの前に姿を現してくれたのは、偶然ではない。私たちの世代に託された警鐘としてウィルソンの写真を見つめ直すことが今、求められている。

2016年5月　　　　古居 智子

Author's Note

Wilson was born when the U.K. was entering the Second Industrial Revolution. It was a time of technological advancements in the fields of chemistry, electricity, oil and steel and the industrial structure shifted from light industries to heavy industries. When he began working in Birmingham, it was an industrial area called the 'black country' where the sun was blocked by smog. When he got a job at Kew Garden, the town of London was suffering serious air pollution from the use of coal fuel. Soseki Natsume, well known Japanese novelist who studied in London at almost same time, wrote in his journal, "Eject the spit from your mouth and see, you will be surprised at a black lump that appears." 100 years later the same situation exists in places like Beijing or New Delhi.

Coming to Japan after having spent 9 years in rugged and remote regions of China, Wilson remarked; "Modernization of Japan is progressing too fast." With the goal of catching up with the Western powers, Japan, which gained momentum by the victories of the Sino-Japanese and the Russo-Japanese Wars, pushed earnestly on the way to industrialization. At the same time, copper mining pollution in Gunma prefecture and cadmium poisoning disease in Toyama prefecture were reported as serious health problems.

Wilson ended his profession as plant hunter after his second trip to Japan. In 1920 he went on an 18-month journey to visit the botanical gardens of Australia, Tasmania, New Zealand, Singapore, Malay Federation, India, Ceylon, Kenya and South Africa. What he saw there was the face of devastating deforestation.

"When you destroy the forests you upset the balance of nature. Besides the aspect of utility and beauty trees play an important part in health. They live on the very gas (carbon dioxide) that is poisonous to the animal kingdom, and if the vegetable kingdom were destroyed the animal kingdom would automatically die, too. The destruction of the great forests to make way for expansion was shortsighted and would ultimately lead to the loss of one of the greatest natural assets: a plentiful supply of valuable timber. The present generation must remember that it only holds the forests in trust for future generations."

Wilson came to speak frequently of concerns about the influence that environmental destruction has on human life. The term global warming was not born yet, but he had already discerned the crisis of the earth by the increase of carbon dioxide 100 years ago.

Can we hold our heads up and say we are any smarter than the people of Wilson's time ? The great problems of abnormality of the weather, the destruction of ecosystems, for which Wilson sounded the alarm-bell, have only intensified.

That Wilson's photographs and words have resurfaced at this turning point of 100 years is no accident. Reflecting on his work in these critical times requires this generation to heed his warning bell.

May, 2016 Tomoko Furui

植物名索引　List of Plants

＊ 標準和名、学名は APGIII 体系および「植物和名ー学名インデックス YList」に則る
Using 'APG III classification' and 'YList, Japanese Plant Name-Scientific Name Index.'

著者紹介　Author's Profile

古居 智子（ふるい・ともこ）
Tomoko Furui

大阪生まれ。北海道大学卒。米国ボストンでジャーナリストとして活躍後、1994 年米国人建築家の夫とともに屋久島に移住。NPO 法人屋久島エコ・フェスタ理事長。環境保護活動に励みながら、屋久島の文化、暮らし、歴史、自然などをテーマに執筆活動を続けている。5 年前からウィルソンの調査を開始。資料の発掘と取材執筆に情熱を注ぐ。

Born in Osaka. Graduated from Hokkaido University. Worked as a journalist in Boston, U.S.A. Moved to Yakushima with her American husband in 1994. Serves as Director of NPO Yakushima Eco-festa. Committed to writing books about culture, life, history, and nature of Yakushima while working on actives to protect the environment. Since 2011 researching the life, work and writings of Ernest Henry Wilson.

著書：『夢みる旅「赤毛のアン」』（文藝春秋）『屋久島 恋泊日記』（南日本新聞社）『屋久島 島・ひと・昔語り』（南日本新聞開発センター）『密行 最後の伴天連シドッティ』（新人物往来社）『はじまりのかたち - 屋久島民具もの語り』（NPO 法人屋久島エコ・フェスタ）『ウィルソンの屋久島 -100 年の記憶の旅路』（KTC 中央出版）〈平成 26 年度南日本出版文化賞受賞〉など

Chief Literary Works：（Japanese）"Travels with Anne, Prince Edward Island" "A Diary of Koidomari, Yakushima" "Yakushima, Islanders and Folklore" "The Last Missionary, Shidotti" "Yakushima's Folk Craft"／（Japanese & English）"Wilson's Yakushima-Memories of the Past"〈Awarded Best Book in Southern Japan, 2014〉

本書は、公益財団法人自然保護助成基金プロ・ナトゥーラ・ファンド助成を受けて出版されました。

ウィルソンが見た鹿児島　プラント・ハンターの足跡を追って
Wilson's Kagoshima　Tracing the Footsteps of a Plant Hunter

2016 年 5 月 20 日　初版第 1 刷発行

著　者　古居智子
発行者　向原祥隆
発行所　株式会社 南方新社
Nanpou Shinsha Kagoshima
〒892-0873　鹿児島市下田町 292-1
電話　099-248-5455
振替口座　02070-3-27929
URL　http://www.nanpou.com/
e-mail　info@nanpou.com

印刷・製本　モリモト印刷株式会社
定価はカバーに表示してあります　乱丁・落丁はお取り替えします
ISBN978-4-86124-337-0 C0045